ENVIRONMENTAL REMEDIATION TECHNOLOGIES, REGULATIONS AND SAFETY

EPA OVERSIGHT OF PESTICIDES

GOOD LABORATORY PRACTICES AND CONDITIONAL REGISTRATIONS

ENVIRONMENTAL REMEDIATION TECHNOLOGIES, REGULATIONS AND SAFETY

Additional books in this series can be found on Nova's website under the Series tab.

Additional e-books in this series can be found on Nova's website under the e-book tab.

ENVIRONMENTAL REMEDIATION TECHNOLOGIES, REGULATIONS AND SAFETY

EPA OVERSIGHT OF PESTICIDES

GOOD LABORATORY PRACTICES AND CONDITIONAL REGISTRATIONS

MAGGIE GIBSON
EDITOR

nova publishers
New York

For permission to use material from this book please contact us:
Telephone 631-231-7269; Fax 631-231-8175
Web Site: http://www.novapublishers.com

NOTICE TO THE READER

Additional color graphics may be available in the e-book version of this book.

Library of Congress Cataloging-in-Publication Data

ISBN: 978-1-63463-177-8

Published by Nova Science Publishers, Inc. † New York

CONTENTS

PREFACE

This book examines the extent to which EPA inspects for GLP compliance laboratories that test pesticides and the challenges EPA faces in doing so; uses the information obtained through GLP inspections in its pesticide decision-making process; and collaborates with FDA on GLP inspections.

Chapter 1 – Pesticides used to control weeds, unwanted insects, and fungi contribute to agricultural productivity and public health by preventing crop damage and controlling pests. However, pesticides may also have adverse effects. EPA's OPP reviews applications for pesticide products and registers those that it determines do not have unreasonable adverse effects on health and the environment. EPA's OECA inspects laboratories where these pesticides are tested to ensure that the laboratories followed EPA's GLP regulations. FDA also conducts GLP laboratory inspections. GAO was asked to review EPA's GLP Compliance Monitoring Program. This report examines the extent to which EPA (1) inspects for GLP compliance laboratories that test pesticides and the challenges EPA faces in doing so, (2) uses the information obtained through GLP inspections in its pesticide decision-making process, and (3) collaborates with FDA on GLP inspections. To conduct this work, GAO reviewed relevant agency documents and data, conducted a nongeneralizeable survey of 20 laboratories, and interviewed EPA and FDA officials and laboratory and other stakeholders.

Chapter 2 – As of September 2010, more than 16,000 pesticides were registered for use in the United States, according to EPA. EPA reviews health and environmental effects data submitted by a company and may register a pesticide or, alternatively, grant a "conditional registration" for a pesticide under certain circumstances, even though some of the required data may not have been submitted or reviewed. The company must provide the missing data within a specified time. In 2010, environmental and other groups charged that EPA had overused conditional registrations and did not appear to have a reliable system to identify whether the required data had been submitted. GAO was asked to examine issues related to EPA's use of conditional registrations for pesticides. This report examines the (1) number of conditional registrations EPA has granted and the basis for these, (2) extent to which EPA ensures that companies submit the required additional data and EPA reviews the data, and (3) views of relevant stakeholders on EPA's use of conditional registrations. GAO reviewed EPA data and surveyed stakeholders, among other things.

In: EPA Oversight of Pesticides
Editor: Maggie Gibson

ISBN: 978-1-63463-177-8

Chapter 1

PESTICIDE SAFETY: IMPROVEMENTS NEEDED IN EPA'S GOOD LABORATORY PRACTICES INSPECTION PROGRAM*

United States Government Accountability Office

WHY GAO DID THIS STUDY

Pesticides used to control weeds, unwanted insects, and fungi contribute to agricultural productivity and public health by preventing crop damage and controlling pests. However, pesticides may also have adverse effects. EPA's OPP reviews applications for pesticide products and registers those that it determines do not have unreasonable adverse effects on health and the environment. EPA's OECA inspects laboratories where these pesticides are tested to ensure that the laboratories followed EPA's GLP regulations. FDA also conducts GLP laboratory inspections. GAO was asked to review EPA's GLP Compliance Monitoring Program. This report examines the extent to which EPA (1) inspects for GLP compliance laboratories that test pesticides and the challenges EPA faces in doing so, (2) uses the information obtained through GLP inspections in its pesticide decision-making process, and (3) collaborates with FDA on GLP inspections. To conduct this work, GAO reviewed relevant agency documents and data, conducted a nongeneralizeable

* This is an edited, reformatted and augmented version of the United States Government Accountability Office publication, GAO-14-289, dated May 2014.

survey of 20 laboratories, and interviewed EPA and FDA officials and laboratory and other stakeholders.

WHAT GAO RECOMMENDS

GAO recommends, among other things, that EPA assess its authority and need for a fee-based inspection system for the GLP Program, determine why the database information to prioritize laboratories is incomplete, and that EPA and FDA develop a process to collaborate and share information on planned and completed inspections. EPA and FDA agreed with GAO's recommendations.

WHAT GAO FOUND

The Environmental Protection Agency (EPA) inspects few laboratories that test pesticides for Good Laboratory Practices (GLP) compliance and faces challenges in selecting laboratories to inspect. For fiscal years 2009 to 2013, EPA's Office of Enforcement and Compliance Assurance (OECA) GLP Compliance Monitoring Program inspected from 51 to 80 laboratories annually from an estimated 1,400 that conducted studies used to support applications for pesticide registrations. During the same period, EPA reduced OECA's GLP Compliance Program budget and staff by about half. Laboratory and other stakeholders told GAO that not having GLP inspections can negatively affect a laboratory's business domestically and abroad. OECA prioritizes laboratories for GLP inspections using criteria that reflect, among other things, how long it has been since the last inspection and the number of studies the laboratory has conducted that have been submitted to EPA's Office of Pesticide Programs (OPP) in support of a pesticide registration application. However, GAO found that some laboratory information in the OECA database used to prioritize inspections was either inaccurate or incomplete, making it difficult to target laboratories for inspections. GAO also found that OECA is considering ways the GLP program could be run more efficiently given its recent budget cuts and concerns of stakeholders about the infrequent GLP inspections. For example, OECA officials have informally discussed the possibility of charging user fees that may be used to fund the GLP program, as the U.S. Food and Drug Administration (FDA) and many other countries do, but the agency has not conducted a formal evaluation of user fees. Without formally assessing the need for such fees, EPA cannot determine whether charging and retaining the fees would be possible and whether such fees could help make the inspection program self-sustaining.

EPA rarely uses GLP inspection results in making its initial pesticide registration decisions. An OPP official said that this is because most inspections occur after decisions have been made. OPP officials said they have not denied or revoked any pesticide registrations based on OECA GLP inspections during the past 5 years, but OPP has taken other actions, such as requiring that a study be repeated because of subsequent laboratory inspection information. According to EPA officials, OPP and OECA have communicated on an informal basis about OPP's inspection priorities before a pesticide registration has taken place.

EPA and FDA do not regularly collaborate on laboratory inspections and may be duplicating each other's work at some of these laboratories. In 1984, EPA and FDA entered into an agreement to collaborate on GLP inspections and met quarterly to discuss upcoming inspections; the agreement ended in 2004 although meetings continued until 2007. From fiscal year 2005 to 2012, EPA and FDA conducted a total of 170 GLP inspections of the same 37 laboratories. In 38 of 170 inspections, the agencies inspected the same laboratory during the same fiscal year. EPA and FDA have independent but similar sets of GLP regulations. Officials from both agencies said it would be useful to know which laboratories the other agency was planning to inspect and to have those inspection results, since each agency can only inspect a certain number of laboratories each year. Absent collaboration and information sharing with FDA on planned and completed GLP inspections, EPA will have difficulty efficiently using its limited resources to increase the number of inspections it conducts.

ABBREVIATIONS

CFO	Chief Financial Officer
EPA	Environmental Protection Agency
FDA	Food and Drug Administration
FFDCA	Federal Food, Drug and Cosmetic Act
FIFRA	Federal Insecticide, Fungicide, and Rodenticide Act
GLP	Good Laboratory Practices
HHS	Department of Health and Human Services

IBT	Industrial Bio-Test Laboratories
ICIS	Integrated Compliance Information System
LISA	Laboratory Information and Study Audit
MAD	Mutual Acceptance of Data
OCE	Office of Civil Enforcement
OCEFT	Office of Criminal Enforcement, Forensics and Training
OECA	Office of Enforcement and Compliance Assurance
OECD	Organisation for Economic Co-operation and Development
OIG	Office of Inspector General
OMB	Office of Management and Budget
OPP	Office of Pesticide Programs
OPPIN	Office of Pesticide Program Information Network
PRIA	Pesticide Registration Improvement Act of 2003
TSCA	Toxic Substances Control Act

***

May 15, 2014

The Honorable Paul D. Tonko
Ranking Member
Subcommittee on Environment and the Economy
Committee on Energy and Commerce
House of Representatives

Dear Mr. Tonko:

Pesticides that are used to destroy or control weeds, unwanted insects, fungi, rodents, bacteria and other pests contribute to agricultural productivity by preventing crop damage and improving public health by controlling disease carrying pests. However, the use of pesticides may also have adverse effects on human health and the environment. For example, some pesticides in large doses may contribute to the risk of cancer and other serious human health problems and may also cause environmental damage, such as contamination of water supplies or unintended impacts on species not targeted by the pesticide. The Federal Insecticide, Fungicide, and Rodenticide Act (FIFRA), as amended,[1] generally requires registration of pesticides and authorizes the Environmental Protection Agency (EPA) to limit the distribution and sale of unregistered pesticides to the extent necessary to prevent unreasonable adverse effects on the environment.[2] Under FIFRA implementing regulations, EPA reviews applications for pesticide products and registers those that it determines will meet the FIFRA statutory standards for registration.[3] EPA also inspects the laboratories where these pesticides are first tested for safety. Laboratories must conduct studies in support of pesticide registration applications in accordance with regulations called Good Laboratory Practices (GLP), which are intended to ensure the quality and integrity of data.[4] GLP regulations were developed and promulgated in the United States during the late 1970s and early 1980s in response to fraudulent laboratory activities and poor laboratory practices that occurred during that time.

To monitor a laboratory's GLP compliance with these standards, EPA conducts both an audit of the studies conducted by the laboratory and submitted to EPA and an on-site inspection of the laboratory; these two types of reviews together are referred to as a GLP inspection.[5] EPA's Office of Pesticide Programs (OPP) reviews studies that support applications for registration of pesticide products submitted by study sponsors and makes registration decisions based on whether or not the pesticide products meet the FIFRA standard for registration.[6] EPA's Office of Enforcement and Compliance Assurance (OECA) has a Compliance Monitoring Program that conducts inspections to ensure that laboratories comply with the agency's GLP regulations. If OECA's inspectors find a problem with the studies they audit[7] or the laboratories they inspect, OECA submits this information to OPP for reexamination of the studies to ensure that the data they generated are valid and reliable to support pesticide decisionmaking. If OPP determines the data upon which a registration decision was based are not valid or reliable, it can ask the manufacturer of the pesticide to repeat a study, or it may initiate steps to change a registration decision.

The Food and Drug Administration (FDA), an agency of the Department of Health and Human Services (HHS), also requires that laboratories follow FDA's GLP regulations, promulgated under the Federal Food, Drug and Cosmetic Act (FFDCA)[8] for conducting nonclinical laboratory studies of investigational drugs, medical devices, food additives, and other products to ensure the quality and integrity of the safety data.[9] In some cases, EPA and FDA receive studies from the same laboratories.

Environmental and other groups have raised concerns about OPP's pesticide registration program, as we reported in August 2013,[10] when we recommended that EPA take steps to improve the program. Pesticide manufacturers have also expressed concern about how infrequently OECA conducts GLP inspections, which they have stated affects firms' ability to register and market their products in the United States and abroad.

You asked us to review EPA's GLP Compliance Monitoring Program. This report examines the extent to which (1) EPA inspects for GLP compliance laboratories that test pesticides and the challenges, if any, EPA faces in doing so; (2) EPA uses the information obtained through GLP laboratory inspections in its pesticide decision-making process; and (3) EPA and FDA collaborate on GLP inspections.

To conduct this work, we reviewed relevant laws, regulations, EPA and FDA guidance, EPA data on pesticide studies, and GLP inspection results. We interviewed OECA and OPP officials, as well as FDA officials. We reviewed recent literature related to GLP, including information and documents found on the websites of a variety of industry, international, environmental, and academic organizations, and foreign government GLP inspection programs. To determine the extent to which EPA inspects for GLP compliance laboratories that test pesticides, we collected and analyzed documentation from EPA officials on its GLP inspection process and analyzed EPA laboratory and inspection data, the agency's use of the data, and the accuracy and completeness of the data. We determined that inspection data were sufficiently reliable to present results on the number of inspections conducted during fiscal year 2009 to fiscal year 2013, based on a comparative analysis of inspection data contained in two of EPA's databases and interviews with agency officials. We determined that data on laboratories' identification numbers and addresses were not sufficiently reliable to assess the number and location of laboratories that had conducted studies submitted to EPA and, as a result, were subject to a GLP inspection. We determined the data were not sufficiently reliable, based on our analysis of the data, identification of erroneous data, and interviews with agency and laboratory representatives. However, EPA did provide estimates of the number and percentage of eligible laboratories that were inspected each year, and we used that information in this report. We obtained opinions on the GLP inspection program, including potential challenges, by interviewing individuals representing 25 entities, including laboratories; pesticide manufacturers; international organizations; environmental and health organizations; national and trade associations; and foreign government GLP programs, which we selected based on referrals from various stakeholders and EPA officials. In addition, we obtained survey responses from 20 laboratories of a nongeneralizable random sample of 53 laboratories and other entities that had sent supporting studies to OPP for a pesticide registration application for fiscal year 2010 to fiscal year 2012 to determine if they had been inspected by

EPA[11] (see app. II). If they had not had a GLP inspection, we asked what impact, if any, they thought not having a GLP inspection had on their business. Although the results of this sample are not generalizable, the laboratories were randomly selected, and the results can illustrate such laboratories' experiences and challenges with GLP inspections. To assess the extent to which OPP uses the information obtained through GLP laboratory inspections in its pesticide registration decision-making process, we analyzed relevant documents and databases to determine the number of OECA's GLP inspections that produced results that were referred to OPP for reexamination, the time it took to conduct these reexaminations, and any impact the inspections had on OPP's pesticide registration decisions. To determine the extent to which EPA and FDA collaborate on inspections, we analyzed EPA and FDA GLP laboratory inspection data. We determined that the inspection data were sufficiently reliable for our purposes. We also analyzed the 20 survey responses to determine if the laboratories had conducted GLP tests or studies for submission to both EPA and FDA since 2008. We reviewed agency documents, such as a 1984 interagency agreement between EPA and FDA to cooperate on GLP inspections, and we interviewed EPA and FDA officials and laboratory representatives about the potential for the two agencies to collaborate. Appendix I contains more detailed information on our objectives, scope, and methodology.

We conducted this performance audit from November 2012 to May 2014 in accordance with generally accepted government auditing standards. Those standards require that we plan and perform the audit to obtain sufficient, appropriate evidence to provide a reasonable basis for our findings and conclusions based on our audit objectives. We believe that the evidence obtained provides a reasonable basis for our findings and conclusions based on our audit objectives.

BACKGROUND

The primary federal laws that govern how EPA regulates pesticides in the United States are FIFRA and FFDCA. Under FIFRA, EPA registers pesticides distributed, used, or sold, in the United States and prescribes labeling and other regulatory requirements to prevent unreasonable adverse effects on health and the environment.[12] If use of a pesticide would result in a residue of the substance in or on food, EPA may not register a pesticide under FIFRA unless, among other things, it can determine that the residue is "safe" as defined by section 408 of the FFDCA.[13] EPA may establish a tolerance level—the maximum permissible pesticide residue in or on food or animal feed that is sold— that meets the FFDCA safety standard set forth in section 408 or may choose to grant an exemption for a tolerance. To obtain a pesticide registration, or petition to establish a tolerance level, a registrant (company or person) submits an application for EPA's review containing health and environmental effects data and other information on a pesticide.[14] In addition to testing conducted in a laboratory, these data can include studies based on the analysis of data collected from field applications of the pesticide and other information on the pesticide. The application is submitted with a pesticide registration fee, with the fee amount depending on certain factors, such as the type of registration action that is being requested. According to an OPP official, registrants are required to provide data relating to the pesticide's hazard and exposure levels to accompany a proposed registration. All studies submitted in support of a product registration are subjected to an initial screening and regulatory and scientific review by OPP staff to determine potential human and environmental risks. OPP reviews the applications and registers those products determined to meet FIFRA's standards for registration and other regulatory requirements so that they can be marketed domestically.

Under FIFRA, when a registrant submits an application for pesticide products to OPP, the accompanying data to support the registration are required to be prepared in accordance with the agency's GLP regulations to ensure the quality and integrity of the data in the study. Each study submitted is reviewed by OPP to see if GLP criteria have been met. OPP staff prepare science reports and/or data evaluation records to address the quality of each study. If studies submitted in support of this process contain questionable data, this can result in adverse action by OPP. For example, according to OPP officials, if OPP finds that a registrant provided false or inaccurate information about the certified limits of the active ingredients of a pesticide, or if there is a discrepancy in the number of formulas used to check concentrations, a study would be rejected, and OPP might deny such an application or request that it be withdrawn.

OECA's GLP Program is tasked with monitoring compliance with the GLP regulations through on-site inspections of laboratories, as well as audits of the laboratories' studies. According to EPA officials, the purpose of the study audit is to ensure that the study was conducted in accordance with the agency's GLP regulations and that the study is supported by data generated by the laboratory. A study's sponsor or applicant must include one of the following statements in the submission: (1) that the study was conducted in accordance with GLP, (2) a detailed description of all the differences between the practices used in the study and those required by GLP, or (3) that they do not know if their study was conducted in accordance with GLP.[15] Laboratories eligible for inspection are those that have conducted studies submitted to the agency. In some instances, a single laboratory may have conducted more than one submitted study. During an inspection, OECA's GLP Program inspectors will verify the accuracy of a study's GLP compliance statement. GLP inspections can take place while OPP is still reviewing a pesticide registration application or a petition for a tolerance-setting, or it can take place after a pesticide has been registered, or the tolerance has been established. Figure 1 summarizes the pesticide registration and tolerance-setting process, as well as its relationship to OECA's inspection of GLP laboratories.

EPA and FDA have each developed their own GLP standards to address problems found with laboratory studies submitted for the agencies' review. Investigations by these agencies in the mid-1970s revealed that some studies had not been conducted in accordance with commonly accepted laboratory practices. For example, according to an industry representative, one of the first laboratories to attract regulatory and media attention was Industrial Bio-Test Laboratories (IBT), a contract toxicological research laboratory that conducted much of the U.S. toxicological testing at the time. As a result of EPA's and FDA's investigations of IBT, several hundred studies were invalidated because of deliberate fraud, and hundreds of chemicals had to be retested. Specific findings included poor recordkeeping, testing conducted by untrained and unqualified personnel, and data fabrication. For example, data were submitted on rats that had previously been reported as deceased. As a result, in 1978, FDA formulated and published GLP regulations under FFDCA.[16] In 1983, EPA published its GLP standards for pesticide toxicology studies, and in 1989, EPA extended the standards' coverage to include nearly all research data supporting pesticide registrations under FIFRA. The Organisation for Economic Co-operation and Development (OECD), which includes more than 30 member countries including the United States, published OECD *Principles of Good Laboratory Practices and OECD Guidelines for the Testing of Chemicals* in 1981.[17]

GLP standards cover the proper handling of laboratory test substances, equipment maintenance and calibration, testing operations, study plans, quality assurance, recordkeeping and reporting requirements, and facility management, among other things. For example, when EPA conducts a GLP inspection, it determines, among other things, whether the laboratory is of suitable size and construction to facilitate the proper conduct of the studies.

Specifically, an EPA inspection would determine, among other things, whether the laboratory has a sufficient number of rooms or other areas for proper separation of species and testing, as well as for the collection and disposal of contaminants and waste.

In 1981, OECD established the Mutual Acceptance of Data program designed to obtain international recognition of testing data in support of pesticide registrations. As of August 2012, there were 31 member countries, including the United States, and 5 nonmember countries, each participating on a rotating basis in evaluating each other's testing programs compared with OECD test guidelines. EPA has signed memorandums of understanding with seven countries under the Mutual Acceptance of Data Program and has a bilateral agreement with China.[18] According to EPA documents, U.S. companies seeking to sell pesticides in foreign countries and multinational companies seeking registration of pesticides to sell in the United States must meet two Mutual Acceptance of Data program criteria: (1) the country where a study was conducted has a valid and active GLP Compliance Monitoring Program and (2) the testing facility (or laboratory) was inspected by the country's GLP compliance monitoring authority. Some other countries fund their GLP inspections through fee-based systems, under which the registrant or laboratory pays a portion or all of the inspection costs. For example, according to a European Commission official, there are currently 17 European Union member countries that charge some type of fee for conducting GLP inspections. According to FDA officials, FDA charges user fees for medical and animal product application reviews; these fees, together with appropriated funds, provide resources that FDA uses for conducting GLP inspections related to such products. According to EPA officials, although the agency charges fees for review of pesticide applications by OPP, it does not charge a fee for OECA's GLP laboratory inspections.

EPA's GLP inspection process includes three stages: (1) the preinspection targeting stage, (2) the inspection stage or on-site inspection of the laboratory, and (3) the postinspection stage.

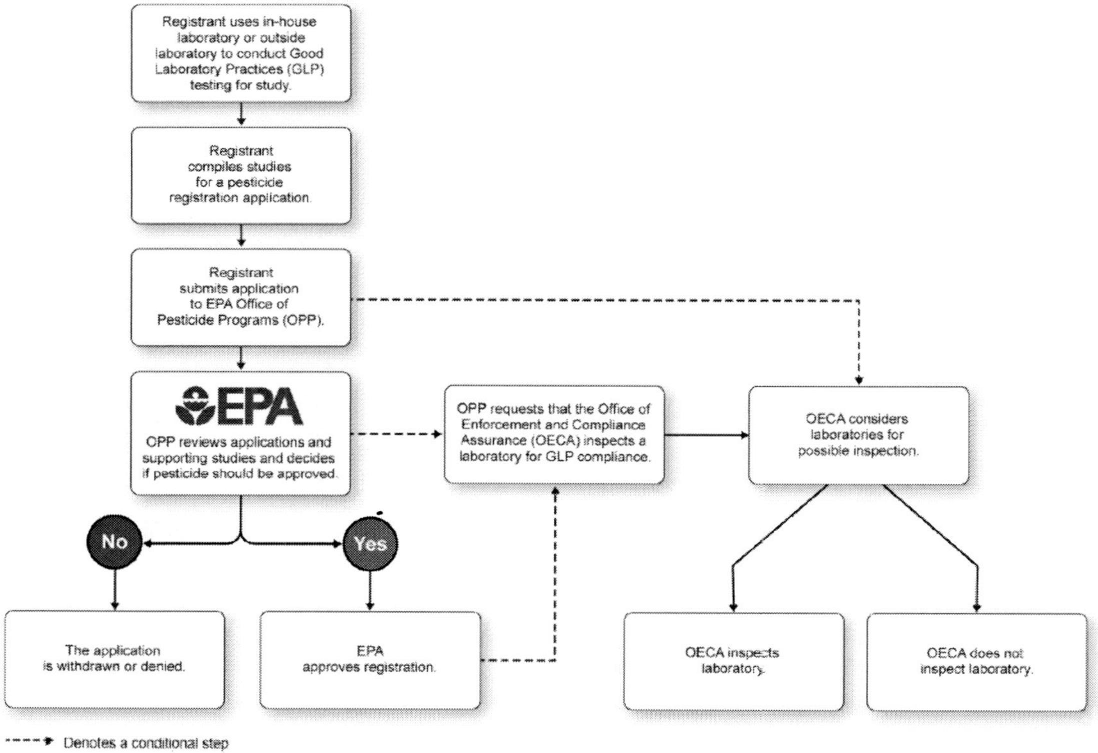

Source: GAO analysis of EPA documents.

Figure 1. The Role of Good Laboratory Practices in the Pesticide Registration Process.

Preinspection Targeting Stage

The preinspection targeting stage includes a series of activities performed by inspectors before the actual on-site inspection is conducted. OECA's GLP Compliance Monitoring Program usually initiates its own inspections by using a targeting module—referred to as the "Neutral Scheme Targeting Module"—that automatically searches the GLP Program's Laboratory Information and Study Audit (LISA) database to target potential studies and pesticide test laboratories. The GLP Compliance Monitoring Program selects laboratories for inspection from among the population of laboratories that have conducted studies submitted to OPP.[19] Using the Neutral Scheme Targeting Module, the inspector identifies laboratories and applies numerical weights to them based on a set of criteria (e.g., the length of time since the last inspection, the severity of the last and prior inspection findings, and the number of submitted studies conducted by the laboratory) to establish a list of potential laboratories to inspect. Most laboratories are selected for inspection through the Neutral Scheme Targeting Module.[20] When laboratories are identified for inspection through use of the Neutral Scheme Targeting Module, the GLP Program Manager assesses these laboratories to determine their eligibility for inspection along with the studies selected for audit. The assessment includes performing a preliminary review of studies to determine, among other things, whether the laboratory is still in business and whether any studies conducted by the laboratory have been rejected by OPP in a review of a pesticide application. The GLP Program Manager will then prioritize and select the laboratories for inspection. Criteria for selection can include whether the eligible laboratories are in the same geographic area. The inspector determines when the laboratory will be inspected and, 10 days prior to the inspection, notifies the laboratory of the impending inspection.

Inspections can also be initiated at OPP's request or that of another EPA office. For example, as OPP reviews studies, it may have questions or problems with the supporting data. In those cases, OPP can request that OECA's GLP Program conduct an audit of particular studies and inspect the laboratories used for these studies. In addition, according to EPA officials, foreign governments can request an EPA inspection if a U.S. registrant is applying to use its pesticide in a foreign country and that country requires a GLP inspection of the U.S. laboratories whose studies are to be used in support of the registration. Finally, according to EPA officials, OECA can conduct a "for cause" GLP inspection based on a tip or complaint received through anonymous phone calls, e-mails or letters about potential GLP violations at a laboratory.

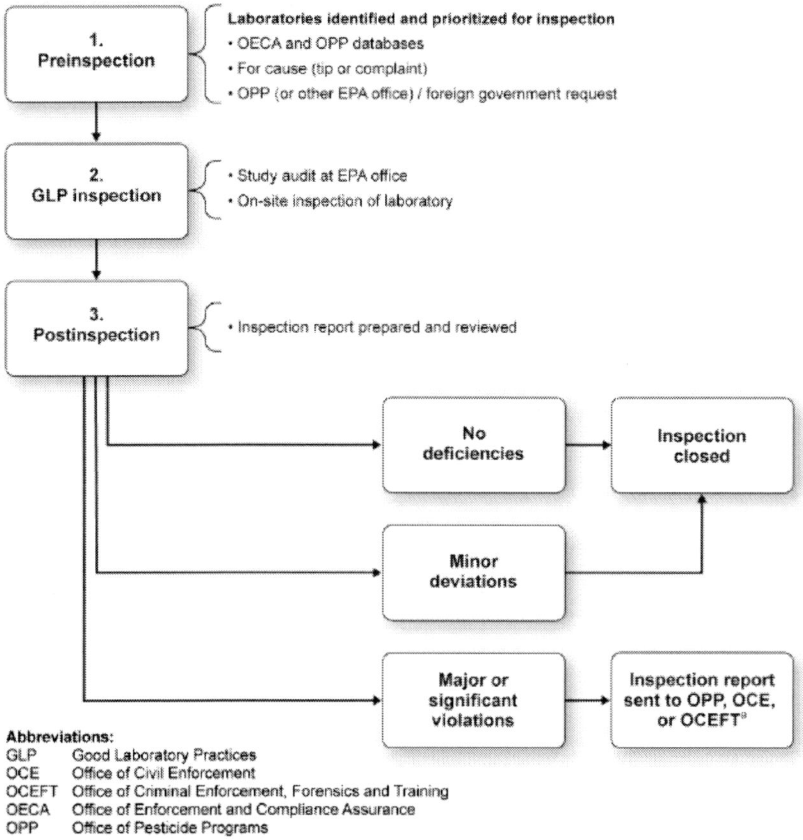

Source: GAO analysis of EPA documents.

[a] The GLP Program Manager will refer significant violations of the GLP standards to OPP and, if appropriate, to OECA's OCE. If a violation is criminal in nature, it will be referred to OECA's OCEFT.

Figure 2. EPA Office of Enforcement and Compliance Assurance (OECA) GLP Compliance Monitoring Program's Inspection Process.

Inspection Stage

The inspection stage begins with the OECA inspectors, prior to visiting a laboratory, conducting study audits and evaluating the laboratory studies submitted to OPP to determine whether they adhere to GLP standards. GLP Compliance Monitoring Program inspectors previously performed this data audit during the inspection of the laboratory. In April 2012, EPA changed its process to have inspectors audit studies before inspecting the laboratory to save time spent at the laboratory and therefore reduce travel expenses. However, OECA officials told us that its inspectors follow the old process if appropriate to the situation. The inspectors are also to verify that data generated by the laboratory support the conclusions made in the study.

Next, an inspector travels to the laboratory to perform the on-site inspection of the laboratory. As part of this inspection, the inspector performs a walk-through of the laboratory to gain an understanding of its capabilities. The inspector will ask to see and review inspection documentation associated with the laboratory, including qualifications and responsibilities of laboratory personnel, the receipt and storage of test substances, the laboratory's standard operating procedures, and animal rooms. The inspector then conducts a compliance inspection to gauge the laboratory's current practices and procedures by selecting one or two ongoing or recently completed studies from the laboratory's master schedule to see whether they adhered to GLP standards. The inspector also observes a procedure in an ongoing study to determine adherence to standard operating procedures, the study plan, and GLP, as well as whether the person performing the procedure has the appropriate education and experience.

Postinspection Stage

The postinspection stage involves activities such as preparing an inspection report that contains a description of the general nature of the inspection and the laboratory being inspected. Report appendixes cover the findings from the study audits and the on-site inspection. Inspection reports are next reviewed by another GLP inspector to ensure the reports are complete and consistent. According to an EPA official, the GLP Program Manager then reviews the inspection report and assigns one of the following categories to the laboratory as specified in the Enforcement Response Policy for FIFRA GLP Regulations: (1) no GLP deficiencies noted, (2) minor GLP deviations, or (3) major/significant GLP violations. For laboratories given ratings in the first two categories, the inspector closes out the inspection and enters all the inspection information and the review category into the LISA database. The inspector also enters some of this information, such as laboratory information, inspection dates, number of study audits, and potential deficiencies, into OECA's Integrated Compliance Information System (ICIS), which is used for OECA-wide compliance reporting results.

The GLP Program Manager sends a letter to the laboratory and the registrant notifying them of the results of the inspection and, in the case of any deficiencies, requesting assurances that appropriate action will be taken to remedy the deficiencies. In the case of a significant violation of the GLP standards that, in the opinion of the inspector, may have affected the validity and integrity of studies performed at the laboratory, the GLP Program Manager will refer this information to OPP for consideration in their decision-making process and, if appropriate, to OECA's Office of Civil Enforcement (OCE), which develops and prosecutes administrative civil and judicial cases and provides legal support for cases and investigations initiated in EPA regions. If the violation is criminal in nature, it will be referred to OECA's Office of Criminal Enforcement, Forensics and Training (OCEFT). The inspection will not be closed until OPP or OECA has made a final determination. Figure 2 summarizes the GLP inspection process.

EPA Has Inspected Few Laboratories for GLP Compliance and Faced Challenges in Selecting Laboratories for Inspections

EPA's OECA has inspected few laboratories on an annual basis that test pesticides for GLP compliance for fiscal year 2009 through fiscal year 2013 and faced challenges in doing so. Because of resource limitations in the GLP Compliance Monitoring Program, OECA prioritizes laboratories for GLP inspections based on a set of criteria. However, our analysis showed that some laboratory information in the databases used to prioritize laboratories for inspection was either inaccurate or incomplete, and these data challenges may negatively affect prioritization. OECA officials said that budget cuts, among other things, have reduced the number of inspectors and the number of GLP inspections that can be performed. Given the concerns of some stakeholders about the infrequent GLP inspections, OECA is considering other approaches, such as charging fees or using third parties to review studies, in order to increase the number of inspections.

OECA Inspects Few Laboratories Annually, but Most Laboratories Are Not Regularly Inspected for GLP Compliance

OECA inspected from 51 to 80 laboratories each year for GLP compliance in fiscal year 2009 to fiscal year 2013, but most laboratories were not regularly inspected for GLP compliance. During this period, OECA inspected about 4 to

6 percent of 1,400 eligible laboratories each year, according to EPA data and officials.[21] In contrast, FDA's GLP compliance program guidance manual states that the program's objective is to inspect all eligible nonclinical laboratories conducting safety studies that are intended to support applications for research or marketing of regulated products approximately every 2 years.[22] Similarly, most member countries under OECD's Mutual Acceptance of Data (MAD) agreement conduct inspections of their laboratories for GLP compliance every 2 years, according to an OECD official we interviewed.

Some laboratory representatives we surveyed stated that their laboratories were not inspected for GLP compliance by EPA from 2008 to 2013. EPA officials told us they inspect as many laboratories as they can, given financial constraints. OECA officials explained that a limited number of inspectors and tight budgets have hindered their ability to perform more inspections. In fiscal year 2013, according to EPA officials, OECA had four GLP inspectors to inspect 1,400 eligible laboratories in the United States. OECA officials acknowledged that their four inspectors cannot inspect all 1,400 eligible laboratories for GLP compliance every 2 years. From fiscal year 2009 to fiscal year 2013, EPA reduced its GLP Compliance Monitoring Program budget and the number of full-time equivalent staff by approximately 50 percent from eight to four inspectors. EPA officials said that the reduction in the number of full-time equivalent staff was due to retirements, an inability to hire new inspectors because of budgetary constraints, and most importantly, according to these officials, the inability to find inspector candidates with the requisite skills. During this same time period, the number of studies conducted by laboratories that were submitted to OPP has remained relatively constant, except for an increase in the number of studies submitted in fiscal year 2010. Estimates of the numbers of studies submitted, inspections conducted, inspection staff and GLP budget data are shown in table 1.

To increase efficiency and to respond to a reduced budget and workforce, as well as to address concerns from industry and the OECD, OECA implemented changes to its GLP inspection procedures in fiscal year 2012 to try to increase the number of GLP inspections from the low of 51 conducted in fiscal year 2011.

Under its revised inspection procedures, agency officials consider laboratories' geographic proximity to each other, as well as other criteria, when selecting laboratories to inspect. For example, in February 2013, the agency inspected five laboratories located in Oregon and, in March 2013, the agency inspected two laboratories in Texas. OECA officials explained that grouping laboratory visits geographically saves travel money and staff time. In addition, they stated that a change in OECA's inspection procedures to allow inspectors to audit studies in EPA offices rather than on-site reduced the amount of time that inspectors spent at a laboratory.

Nevertheless, some stakeholders and international officials we interviewed expressed concern about the infrequent GLP inspections by EPA. Pesticide manufacturers and industry associations told us that if laboratories cannot prove GLP compliance to potential clients or to countries where their products may be used, their business can be negatively affected. According to a senior staff person at one U.S. laboratory, in 2011, the Netherlands rejected a study from a U.S.-based laboratory because the contractor laboratory that conducted the study had not been inspected by EPA. This same study had been previously submitted to EPA.

Following the Netherlands' rejection of this study, the laboratory had to repeat the study and subsequently contracted with a European-based laboratory to do so. In addition, two stakeholders we interviewed said that such rejections by other countries could likely increase if U.S laboratories could not show that they had a recent GLP inspection. One laboratory representative said his laboratory may stop doing GLP testing since many registrants require proof of GLP inspections, and his laboratory had not been inspected by EPA. In addition, this laboratory representative said that if U.S. laboratories cannot show that they have had a GLP inspection, registrants may not hire them to conduct studies because of the possibility that those studies may be rejected.

In our survey of laboratories, 5 of 14 respondents reported that EPA could improve its implementation of its GLP Compliance Monitoring Program by increasing the number and frequency of inspections to prevent these business consequences.[23] Twelve of 15 respondents from laboratories with study data submitted to OPP since fiscal year 2008 reported that EPA's GLP inspections had a positive effect on their business. Nine of the 12 respondents who reported positive effects of a GLP inspection indicated that the inspections enabled them to prove GLP compliance to their customers, such as U.S. and international pesticide manufacturing companies, as well as to other countries' governments.

Table 1. Number of Pesticide Studies Submitted, Inspections, Full-Time Equivalents, and Budget for EPA's Good Laboratory Practices (GLP) Compliance Monitoring Program from Fiscal Year 2009 to Fiscal Year 2013

	FY 2009	FY 2010	FY 2011	FY 2012	FY 2013
Number of pesticide studies submitted to EPA[a]	7,791	10,093	8,715	8,326	7,123
Number of GLP inspections[b]	64[c]	71[c]	51[c]	80[c]	79[c]
Number of GLP compliance monitoring program full-time equivalents	8	8	5	4	4
GLP program budget	$1,415,000	$1,021,400	$959,500	$734,500	$702,000

Source: GAO analysis of EPA data.

[a] Multiple studies can, and do, come from the same laboratory.

[b] These numbers include inspections performed under the Federal Insecticide, Fungicide, and Rodenticide Act and exclude inspections conducted under the Toxic Substances Control Act.

[c] Inspection data were taken from EPA's Integrated Compliance Information System database because of technical problems EPA has been experiencing with the Laboratory Information and Study Audit database since July 2011.

OECA Prioritizes Laboratories for GLP Inspections, but Data Challenges May Negatively Affect Its Targeting Process

OECA prioritizes laboratories for GLP inspections, but challenges with some of the data used to determine those priorities may negatively affect its targeting process. GLP Compliance Monitoring Program inspectors use the OPPIN and LISA databases to develop a list of the highest priority laboratories to inspect for GLP compliance in a given year. Once a list is generated using the Neutral Scheme Targeting Module located in the LISA database, OECA officials said they try to select laboratories that are geographically near to one another to inspect. The information used to develop this priority list is based on data transferred from OPPIN to LISA. However, our analysis showed that the laboratories' information in these databases was sometimes incomplete or inaccurate.[24]

We found that the OPPIN system currently has three categories to classify an entity's role in a study: sponsor, performing laboratory, or both. These categories do not allow EPA staff to differentiate between GLP laboratories and other entities involved in a study. As a result, we found some entities were identified in the OPPIN database as "performing laboratories," which EPA defines as facilities where a study or part of a study is conducted, when in fact they were not.[25] For example, of the respondents to our survey of performing laboratories found in the OPPIN database, six reported that they were not a laboratory and did not conduct GLP testing.[26] Of these six, three said they were consultants, two said they contract with an outside source for laboratory testing, and one was a pesticide manufacturer. We were not able to determine how or why these entities were identified as performing laboratories in OPPIN. EPA officials told us that if the entity that submits the study puts inaccurate information on a study's title page concerning who performed the study and where the study was performed, the information entered by EPA's contractors into OPPIN will also be inaccurate. EPA's documentation describing the Neutral Scheme Targeting of laboratories states that the entity's role in a pesticide study determines whether it is a candidate for inspection. If entities are inaccurately identified in OPPIN as performing laboratories, the Neutral Scheme Targeting Module will not consider the correct universe of laboratories to inspect and, as a result, inspectors need to do additional manual research to verify that the laboratory selected for an inspection is indeed an entity eligible for a GLP inspection.

We also found that address information for some laboratories was missing or inaccurate in OPPIN. For example, we found missing address information for 23 percent of laboratories listed in OPPIN.[27] In 2008, EPA updated its 1991 guidance to the contractors responsible for entering data into OPPIN.[28] The updated guidance states that the agency wants to capture the physical address for each laboratory because OECA uses this information to select laboratories for inspection. The guidance acknowledges that determining the performing laboratory in published studies is challenging, but the goal is to identify where the actual location of the research was done. Without accurate address information, there could be problems determining where the testing was actually conducted.

In addition, we found that some laboratory identification numbers were not reliable. In its November 1991 guidance, OPP states that laboratories should be assigned a unique laboratory identification number.[29] For example, we found that one laboratory that was given a ranking in fiscal year 2013 as second highest priority for a GLP inspection had an identification number that was the same as four other laboratories in four different states, making it unclear which of those five laboratories was actually the one selected for inspection.

The agency maintains two different databases to track GLP inspections, ICIS and LISA. The LISA database is used by the GLP Compliance Monitoring Program, while the ICIS database is used EPA-wide, according to officials. Because of technical problems EPA has been experiencing with the LISA database, inspection data have been entered only into ICIS since July 2011, according to officials. We reviewed inspection data contained in both databases for fiscal year 2008 to fiscal year 2013 in an effort to determine the total number of GLP inspections OECA conducted, but we were not able to match the inspection information contained in the two databases. While LISA uses laboratory identification numbers, ICIS does not. As a result, OECA may not be able to track if, or when, a specific laboratory had been inspected because the systems do not both use identification numbers. In addition, the laboratory names are not always the same in the two databases. We were also unable to match up the inspections in the two databases. Without accurate data on which laboratories have been inspected, OECA's Neutral Scheme Targeting Module scoring of laboratories most in need of inspection may be inaccurate and, therefore, manual research is required to ensure that OECA is targeting the laboratories most in need of inspection.

Based on the incomplete or inaccurate data in ICIS, LISA, and OPPIN, it is not always possible for GLP Compliance Monitoring inspectors to identify where the testing was conducted without making telephone calls to individual laboratories. The data issues also impede the agency's ability to efficiently track which laboratories need inspecting. The officials said GLP inspectors must spend time manually verifying laboratory data before deciding which laboratories to inspect. In the absence of reliable data, EPA may not have the data it needs to prioritize laboratories to inspect efficiently or effectively.

OECAIs Considering Other Implementation Approaches to Increase Inspections

OECA is considering other approaches to address laboratory representatives' and others' concerns about infrequent GLP inspections, such as using third parties or charging fees to increase the number of inspections. In August 2012, OECA developed a Budget Adjustment Plan, which provided general information on potential future approaches to the GLP Compliance Monitoring Program, including ways the program could be run more efficiently given its recent budget cuts and inability to hire GLP inspectors. In its Budget Adjustment Plan, EPA states that OECA will evaluate other implementation approaches, such as the use of a third party to conduct GLP inspections. EPA officials told us that they have also conducted internal and informal discussions regarding the possibility of using user fees for the GLP program, but the agency has not conducted a formal evaluation of user fees. Without formally assessing the need for such fees, EPA cannot determine whether charging and retaining the fees would be possible and whether such fees could help make the inspection program self-sustaining.

In 2012, we concluded that federal user fees and charges are generally related to some voluntary transaction or request for government goods or services beyond what is normally available to the public, such as fees for patent applications and customs inspections.[30] In 2012, the federal government collected nearly $300 billion in user fees from the public.[31] In 2008, we found that well-designed user fees can reduce the burden on taxpayers to finance those portions of activities that provide benefits to identifiable users.[32] As we mentioned earlier, FDA charges user fees for medical and animal product application reviews, and these fees, together with appropriated funds, provide resources that FDA uses for conducting GLP inspections related to such products.

In addition, some other countries, such as some that are members of the OECD, fund their GLP inspections through fee-based systems, whereby the registrant or laboratory pays a portion or all of the inspection costs.

According to Office of Management and Budget (OMB) Circular A-25, every 2 years, agencies should review programs that are not currently funded by user fees (such as the GLP program) to determine whether fees should in fact be assessed for government services.[33] Once user fees are implemented, revenue from the fees will be credited to the general fund of the U.S. Treasury as miscellaneous receipts unless otherwise specified by law.[34] Circular A-25 also states that it may be appropriate for an agency to request authority to retain the fee revenue if the user fees offset the expenses of a service that is intended to be self-sustaining. The OMB guidance states that agencies are to discuss the results of the biennial fee reviews and any resulting proposals in their Chief Financial Officers (CFO) Annual Report required by the Chief Financial Officers Act of 1990.[35] In its most recent CFO Annual Report (EPA Agency Financial Report) EPA discussed its biennial review of its existing user fee programs.

However, it did not discuss reviewing the GLP program—or any of its other programs that are not currently funded by fees—to determine whether fees should be assessed.[36] Moreover, in March 2014, the EPA Office of Inspector General (OIG) reported that EPA did not conduct thorough biennial user fee reviews for fiscal year 2008 to fiscal year 2009 and for fiscal years 2010 to 2011, and it did not review all agency programs to determine whether they should assess fees for government services they provided.[37] Among the OIG's recommendations were that the EPA CFO discuss biennial user fee results in the EPA Agency Financial Report, coordinate with programs that claimed an exception to charging fees and costs and help determine whether fees should be assessed. Without assessing the need for user fees, EPA cannot determine whether fees could help make the laboratory inspection program self-sustaining.[38]

When asked about user fees, representatives of 8 of 12 laboratories and pesticide manufacturers told us they would support EPA user fees for GLP inspections, and they cited more frequent GLP inspections as a possible benefit of such user fees. Representatives from the remaining 4 laboratories did not support user fees, and some of these expressed concern that a user fee system might be burdensome to smaller laboratories. However, as we previously reported, the amount of a user fee reflects the cost of providing the service, which may differ among users.[39]

EPA Rarely Used GLP Inspection Results in Initial Pesticide Registration Decisions but Sometimes Used Them for Later Reexamination

EPA rarely used GLP inspection results in making its initial pesticide registration decisions. An OPP official told us that this is because most inspections occur after those decisions have been made. According to an OPP official, pesticide registration decisions are required by statute to be made within 3 to 24 months of receipt of the pesticide application. For example, new pesticide products that are identical in their uses and formulation to one or more products already registered usually require a registration decision within 3 months.[40] However, the official stated that the GLP Monitoring Program staff cannot set up and conduct inspections of many of the laboratories involved within those time frames.

OPP and OECA officials told us that they have historically communicated on an informal basis about OPP's inspection priorities before a pesticide registration has taken place. These priorities are usually based on OPP's review of studies and the concerns that may arise from these reviews or about the sponsor or study performer. This informal communication generally involves OPP contacting OECA by telephone or e-mail to request that certain laboratories be inspected. An OPP official told us that the two offices have recently tried to make better use of the GLP inspection program in OPP's decision making. For example, according to EPA officials, in September 2012, a senior OPP official hand-delivered a document to her OECA counterpart requesting that (1) several laboratories be considered as priorities for GLP inspections in fiscal year 2013; (2) OECA focus its audit of studies on those submitted within the last 6 to 12 months when performing inspections under the Neutral Scheme Targeting Module in fiscal year 2013; and (3) OECA audit specific categories of studies that OPP officials considered critical to their pesticide approval decisions. An OECA official told us that as of March 2014, OECA inspectors had completed all of the eight OPP-requested inspections. An OPP official told us that OPP would like to continue to provide input to OECA prior to the beginning of each fiscal year to increase the opportunity for OPP to react to findings before a product is registered. The two offices appear to be communicating and prioritizing laboratories for GLP inspections informally, but according to an EPA official, there are no documented procedures that define the responsibilities of each office in coordinating and prioritizing GLP inspections. EPA has stated that formal procedures are not needed and that the current method of communication and coordination is satisfactory. However, by relying on informal coordination mechanisms, OECA and OPP are depending on relationships with individual officials to ensure effective coordination, and these informal relationships could end once personnel turnover occurs. Without documented procedures for effective coordination between the two offices, there is no assurance that the two offices will consistently coordinate on GLP inspections in the future. Under federal standards for internal control, agencies are to clearly document internal controls, and the documentation is to appear in management directives, administrative policies, or operating manuals.[41]

When OECA's GLP inspectors find deficiencies at a laboratory, they share the inspection reports with major or significant violations with OPP. According to OPP officials, once OPP receives the OECA inspection report, it sends it to the OPP registration division responsible for the registration (or pending registration) of products being supported by the studies audited. The OPP registration division then reviews the report and conducts a scientific reexamination of the studies that the laboratory in question conducted and that were submitted in support of the pesticide registration. Once OPP finishes its reexamination, if it finds that deficiencies affect a study's findings, OPP is to reject the study and may request that the registrant repeat the study, submit other materials that support that aspect of the registration, or perform other corrective action deemed satisfactory to OPP. After the reexamination is complete, OPP informs both OECA's GLP Compliance Monitoring Program and the registrant of the outcome.

Table 2. Actions by EPA's Office of Pesticide Programs Based on Good Laboratory Practices (GLP) Inspection Results, Fiscal Year 2008 to Fiscal Year 2012

Action	Number of Studies
Requested study be repeated	26
Rejected study and sponsor voluntarily cancelled or voluntarily suspended registration	4
No action needed	20
Review of study still pending	23
Total	73

Source: GAO analysis of EPA data.

Table 3. Average Time from EPA's Office of Compliance and Assurance's (OECA) Good Laboratory Practices (GLP) Inspection of Laboratory to EPA's Office of Pesticide Program's (OPP) Review of Inspection's Impact on Laboratory's Study, Fiscal Year 2008 to Fiscal Year 2012

		Average days	Average months[a]
Completed reviews (of 50 studies)	Average time from start date of OECA's GLP inspection of a laboratory to referral of inspection results to OPP	352	11
	Average time from date OECA referred inspection results to OPP to when OPP completed its review of inspection results' impact on laboratory's study	446	14
	Average time from start date of OECA's inspection to date OPP completed review of study	798	26
Reviews still pending (of 23 studies)	For those reviews that are still pending, average time from date of OECA GLP inspection to March 24, 2014	1066	34

Source: GAO analysis of EPA data.

Note: Totals may not sum due to rounding.

[a] Average months were determined by dividing the average days by 31.

We found that, from fiscal year 2008 to fiscal year 2012, OECA GLP inspectors referred for reexamination the inspection results from 26 laboratories and 73 studies to OPP. Table 2 lists the action taken by OPP as a result of its reexamination of studies, as of March 2014.

Our analysis of EPA data shows that this review process, from the initial OECA GLP inspection to completion of OPP's reexamination, takes on average about 2 years (see table 3).

According to OPP officials, OPP has not denied a pesticide registration or revoked any registrations based on OECA laboratory inspection information during the past 5 years, but OPP has taken other actions because of that information, such as requiring a registrant to repeat a study or requesting that a registrant voluntarily cancel its registration. For example, according to data provided by an OPP official, in fiscal year 2012, 15 studies associated with an already registered pesticide product were found to be "unacceptable" based on

OECA inspection information. For 10 of these studies, EPA informed the registrants that they needed to repeat the study. For the 5 others, the registrant voluntarily canceled or suspended its registration associated with the study, or the study was not a deciding factor in the products registration and therefore EPA took no further action. In addition, OPP has required that an efficacy claim be removed from the label of a registered product based on a study submitted by a laboratory not meeting GLP standards.

OPP officials explained that a single study is not likely to affect the denial or approval of a pesticide registration because OPP usually bases its registration decisions on more than one study (although OPP officials noted their decision depends on, among other things, the type of study). For example, according to OPP officials, some pesticides are the subject of more than 30 different studies before a decision is made on registration. However, these same officials stated that OPP believes that the GLP inspections are valuable not only to alert OPP to issues with study data and laboratories but to act as a deterrent to guard against the improper conduct of studies and submission of fraudulent or incorrect study data. According to OECA officials, GLP inspections often lead to positive action being taken by laboratories and pesticide registrants. They said that registrants have voluntarily withdrawn studies from OPP as a direct result of inspection notifications and inspection findings.

EPA and FDA Do Not Regularly Collaborate on Laboratory Inspections and May Be Duplicating Each Other's Work

EPA and FDA do not regularly collaborate on GLP inspections and may be duplicating each other's work by inspecting the same laboratories.[42] In 1984, EPA and FDA entered into an interagency agreement to collaborate on GLP inspections that was last renewed in 2004. Under the agreement, the two agencies agreed to collaborate in monitoring testing laboratories' adherence to GLP regulations, as well as in auditing of health-related toxicological test reports and related laboratory records. The agencies also agreed to exchange information and coordinate actions concerning active investigations, regulation correspondence, and legal or administrative action being considered against any laboratory covered under the agreement. Among FDA's responsibilities under the agreement was to conduct a certain number of on-site inspections of laboratories identified by EPA. From 2000 to 2007, FDA conducted a total of nine inspections identified by EPA (three in 2000; three in 2001; and one each in 2002, 2005, and 2007).[43] EPA officials said they also formally met with FDA officials on a quarterly and annual basis to discuss upcoming inspections, but that communication ended by 2007 when, according to EPA officials, FDA began selecting laboratories for inspection on an annual basis instead of on a quarterly basis as EPA does. FDA also shifted from having its Office of Regulatory Affairs select laboratories for inspection to a more decentralized system, whereby its centers select laboratories for inspection, which made it more difficult for EPA to learn which laboratories FDA plans to inspect because it has to coordinate with multiple centers instead of the Office of Regulatory Affairs, according to EPA officials.[44] EPA officials also said the last year FDA sent EPA a list of laboratories it planned to inspect was in 2007. They explained that, in the past, when EPA knew which laboratories FDA was going to inspect, it would ask FDA to audit a study for them and would avoid going to the same laboratory in the same year as FDA. However, since these meetings ended, EPA officials said that they do not always know if a laboratory was already inspected by FDA until they arrive on-site. If EPA knew in advance that a laboratory was inspected by FDA recently, EPA inspectors could use FDA's inspection results to inform their decision regarding whether conducting their own inspection was necessary. If FDA's inspection results were sufficient for EPA's purposes, EPA inspectors could potentially select a different laboratory to inspect in place of the one recently inspected by FDA, thereby extending limited resources. It is important to note, however, that, in some circumstances, it may be necessary for both agencies to inspect the same laboratory.

EPA officials stated that they occasionally coordinate with one of FDA's centers on GLP inspections and stated that the two agencies performed a joint GLP inspection in 2013. In addition, EPA and FDA officials also said that both agencies participate in an FDA-led work group focusing on the modernization of laboratories. Officials from both agencies said that it would be useful to know which laboratories the other agency was planning to inspect and to have the results from those inspections. Since each agency only inspects a certain number of laboratories each year, sharing such information could help both agencies leverage resources. This coordination also could increase the number of laboratories that are inspected for GLP compliance, which in turn would help ensure that the study data submitted for pesticide registrations were generated in accordance with GLP regulations. However, the agencies do not regularly collaborate on or communicate about future inspections or share results from completed inspections.

FDA conducts two types of laboratory inspections, surveillance inspections, which are periodic, routine determinations of a laboratory's compliance with GLP regulations and include a facility inspection and study audit, and directed inspections, which are inspections assigned to achieve a specific purpose, such as verifying the reliability, integrity, and compliance of critical safety studies. Because the two agencies do not regularly share GLP inspection-related information, we found that EPA and FDA may be duplicating each other's work in some of their GLP inspections. For fiscal year 2005 to fiscal year 2012, EPA and FDA conducted a total of 170 GLP inspections of the same 37 laboratories.[45] In 38 of the 170 inspections, the agencies inspected the same laboratory during the same fiscal year (see app. III.). For example, EPA inspected a Colorado laboratory in November 2011, and FDA conducted a surveillance inspection of this same laboratory 7 months later in June 2012. Similarly, although EPA inspected a Utah laboratory in January 2012, FDA conducted a surveillance inspection of this same laboratory 5 months later, in June 2012. According to EPA and FDA officials, the GLP standards of the two agencies upon which these inspections are based are largely similar. Moreover, a senior official in OECA's GLP Compliance Monitoring Program and representatives from three laboratories that were inspected by both EPA and FDA told us that the inspections were comparable.

We also found that there is some degree of overlap in the laboratories that are eligible for inspection by the two agencies.[46] Some laboratories covered by GLP regulations conduct tests yielding data that will be submitted to only EPA, but we found that other laboratories conduct tests for review by both EPA and FDA and are, therefore, eligible for inspection by both agencies. FDA officials told us that they did not have data to identify the total number of laboratories that submit data to both agencies and would, therefore, be subject to GLP inspections by both agencies. EPA officials and stakeholders we interviewed, however, told us that laboratories that conduct toxicology testing are the most likely to perform tests that are submitted to both EPA and FDA. For example, one representative from a laboratory in Maryland that had been inspected by both EPA and FDA eight times from fiscal year 2005 to fiscal year 2012 told us that some of the information in the laboratory's toxicology studies FDA officials examined during a 2011 GLP inspection could have been shared with EPA officials. In addition, in our survey, 5 of 19 respondents indicated that their facility conducted tests or studies intended for review by both EPA and FDA. FDA officials told us that there has been no recent communication between the two agencies on which laboratories they plan to inspect and what they may have found at inspections that were conducted, although they have communicated this information in the past. FDA officials said they would welcome a list of inspections planned and conducted by EPA.

EPA and FDA officials said that, at present, they do not share the results of completed inspections at facilities that do both EPA and FDA-related studies, and they do not have a process in place to collaborate on future GLP inspections. As a result, EPA is not learning about laboratories that FDA has inspected and at which it may have found deficiencies. Similarly, FDA is not learning about the results of EPA laboratory inspections. In addition to potentially duplicating each others' work, by not collaborating, the two agencies are missing opportunities to leverage each other's resources and expand their inspection coverage. FDA officials told us it would be helpful to know if EPA had inspected a laboratory and that it would be particularly useful if EPA shared information if the agency had found problems during its inspection. EPA officials also said collaborating and communicating on inspections would be helpful.

In April 2013, we concluded that when executive branch agencies carry out activities in a fragmented and uncoordinated way, the resulting patchwork of programs can waste scarce funds, confuse and frustrate program customers, and limit the overall effectiveness of the federal effort.[47] The federal government uses a range of mechanisms to implement interagency collaboration, such as interagency groups, and interagency agreements and memorandums of understanding.[48] Key practices state that agencies that articulate their agreements in formal documents can strengthen their commitment to working collaboratively.[49] Written agreements are most effective when they are regularly updated and monitored. As we concluded in April 2013, where federal programs or activities are fragmented, overlapping, or duplicative, there are opportunities for agencies to improve the efficiency and effectiveness of government programs and activities.[50] Moreover, without leveraging its inspection resources with FDA, EPA may continue to have difficulty increasing annual GLP inspections of laboratories.

CONCLUSION

OECA has taken some steps to increase the number of GLP compliance inspections of laboratories conducting studies submitted to EPA for pesticide registration. However, OECA officials acknowledge that their four inspectors cannot inspect all 1,400 eligible laboratories for GLP compliance every 2 years, a time frame used by many of the countries that are members of OECD's MAD agreement. Because some laboratories have never had a GLP inspection, laboratory and industry representatives expressed concern that if OECA continues to inspect so few laboratories for GLP compliance, U.S. laboratories and manufacturers will be less competitive with foreign laboratories and manufacturers. In addition, with so many laboratories going uninspected, EPA may not have full assurance of the quality and integrity of data used to make pesticide registration decisions.

OECA officials told us that they have informally discussed the possibility of instituting user fees for the GLP program, but EPA has not conducted a formal evaluation of GLP user fees as directed by OMB guidance for programs not currently funded by such fees and, as recommended by EPA's OIG. OECA might be able to increase the number of inspections it conducts if it were to charge a fee, which may be used to fund GLP inspections, as FDA does, and as do 17 European countries. Without assessing its authority and need for user fees, EPA cannot determine whether fees could make the laboratory inspection program self-sustaining. In addition, we found that the information in EPA databases used by OECA to set priorities for laboratory inspection is sometimes incomplete or inaccurate. In the absence of reliable data, EPA may not have the data it needs to prioritize laboratories to inspect efficiently or effectively.

We commend OECA and OPP staff for communicating with each other on GLP inspection matters, including OECA consideration of OPP's recent steps to identify laboratories that OPP considers to be priorities for GLP inspections. Considering OPP's request is important because most OECA inspections currently take place after OPP has made its registration decisions. However, EPA does not have documented procedures that define the responsibilities of each office in coordinating and prioritizing laboratories for GLP inspections consistent with federal standards of internal control. Without such procedures, there is no assurance that the two offices will consistently coordinate on GLP inspections in the future.

Furthermore, because EPA and FDA do not regularly share inspection-related information, as they did when they had an agreement to collaborate on GLP inspections, the agencies have inspected some of the same laboratories since 2007, while other laboratories may have gone without needed inspections. Without collaboration and information-sharing on planned and completed GLP inspections, EPA and FDA may duplicate GLP inspections, and EPA will have difficulty efficiently using its limited resources to increase the number of inspections it conducts. Officials from both agencies said that collaborating and communicating on inspections would be helpful. However, absent a formal written agreement, it is not clear that the agencies would regularly collaborate on future planned inspections and share results from completed inspections.

RECOMMENDATIONS FOR EXECUTIVE ACTION

We are making four recommendations in this report.

To improve the OECA GLP inspection process, we recommend that the EPA Administrator take the following three steps:

1. Assess the authority and need for a fee-based inspection system, and if such a system is warranted, establish a user fee system, seeking additional legislative authority, if necessary, to make the laboratory inspection program self-sustaining.
2. Direct OECA and OPP to ascertain the exact causes of inaccurate and incomplete data in its databases and take action to ensure that the data, such as identification of performing laboratories and inspection history, are accurately recorded.
3. Direct OECA and OPP to develop documented procedures to coordinate and prioritize laboratories for inspections.

In addition, we recommend that the EPA Administrator and the FDA Commissioner develop a formal written agreement, such as a memorandum of understanding, that outlines how the two agencies plan to regularly collaborate and share information on GLP inspections and avoid duplication of inspections so that EPA can more efficiently use its limited resources.

AGENCY COMMENTS AND OUR EVALUATION

We provided EPA and HHS with a draft of this report for their review and comment; EPA and HHS provided written comments.

In its written comments, EPA stated that it agreed with our findings, conclusions, and all four of our recommendations. In response to our first three recommendations, EPA agreed

- to assess the authority and need for a fee-based GLP inspection system;
- to ascertain the exact causes of inaccurate and incomplete data and ensure that the data are accurately recorded; and,
- that OECA and OPP should develop written procedures to coordinate and prioritize GLP inspections.

In response to our fourth recommendation, EPA agreed to develop written procedures that outline how EPA and FDA will collaborate and share information on GLP inspections. EPA stated that it did not agree that a formal memorandum of understanding between the two agencies was necessary. Our recommendation did not prescribe the type of agreement the agencies should undertake and offered a memorandum of understanding as one example. We agree that written procedures developed and agreed to by both EPA and FDA will address the recommendation.

In its technical comments, EPA noted that, while EPA and FDA have similar GLP regulations, and in a small number of cases both agencies may inspect the same laboratory, EPA does not believe the work is duplicative. However, we define duplication in this report as occurring when two or more agencies or programs engaged in the same activities or provided the same services to the same beneficiaries. We continue to believe that there is a possibility of duplication when both EPA and FDA are inspecting the same laboratories within a short period of time because both agencies provide a similar service to the same beneficiaries (ensuring laboratory compliance with GLP through facility inspections and study audits). Also in its technical comments, EPA acknowledged the problems we found with the databases used for targeting GLP inspections but stated that it does not believe these problems negatively affect targeting for inspection because OECA is able to gather the necessary information manually. While we agree that, ultimately, EPA may target the appropriate laboratories for inspection through its manual research when the databases are inaccurate or incomplete, we believe that the agency's reliance on manual research is not as efficient or effective as using databases containing accurate information.

In its written comments, HHS also agreed with our fourth recommendation. HHS also reiterated the point that there are legitimate reasons why GLP inspections may be conducted by both EPA and FDA at a single laboratory within a relatively short period of time. For example, when FDA observes significant violations during for cause or surveillance inspections, follow-up inspections may be required to verify corrective actions. HHS also provided technical comments on our report, which we incorporated as appropriate.

Sincerely yours,

John Neumann
Acting Director, Natural Resources and Environment

APPENDIX I: OBJECTIVES, SCOPE AND METHODOLOGY

Our objectives were to examine the extent to which (1) the Environmental Protection Agency (EPA) inspects for GLP compliance laboratories that test pesticides and the challenges, if any, EPA faces in doing so, (2) EPA uses the information obtained through Good Laboratory Practices (GLP) laboratory inspections in its pesticide decision-making process, and (3) EPA and the U.S Food and Drug Administration (FDA) collaborate on GLP inspections. To address these objectives, we reviewed relevant federal statutes and regulations,[51] EPA program and guidance documents,[52] federal internal control standards,[53] Office of Management and Budget circulars[54], and previous GAO and EPA Inspector General reports.[55] We also reviewed EPA's fiscal year 2011-2015 strategic plan; EPA's fiscal year 2011, 2012, and 2013 annual agency financial reports; and EPA Office of Enforcement and Compliance Assurance (OECA) fiscal year 2012 Budget Adjustment Plan. In addition, we interviewed EPA's Office of Pesticide Programs (OPP) and OECA officials and reviewed documentation they provided to obtain further information and clarification on EPA's pesticide registration process and how it relates to the GLP process, and we interviewed FDA officials and reviewed documentation they provided on FDA's GLP process. Furthermore, we reviewed recent literature related to GLP, including information and documents found on the websites of a variety of industry, international, environment, and academia organizations, and foreign government GLP inspection programs. We interviewed about 25 representatives from these organizations. We selected these individuals based on referrals from EPA, and industry and environmental stakeholders.

To examine to what extent EPA inspects for GLP compliance laboratories that test pesticides; and the challenges, if any, EPA faces in doing so, we collected and analyzed documentation from OECA officials on its GLP inspection process and analyzed EPA laboratory and inspection data, the agency's use of this data , and the accuracy and completeness of the data. Specifically, we obtained information on how OECA determines which laboratories to inspect and how EPA's Office of Pesticide Programs Information Network (OPPIN), Laboratory Information and Study Audit (LISA), and Integrated Compliance Information System (ICIS) databases are used to assist inspectors in making these decisions, as well as the type of inspection data that are entered into these databases. Regarding inspection data, we requested OECA to provide us with several data elements contained in its LISA database for the years 2000 to 2013, including name and location of laboratory inspected, reason for inspection, number and type of inspections conducted, and number and name of studies associated with the laboratory inspected. We determined that inspection data were sufficiently reliable to present results on the number of inspections from fiscal year 2009 to fiscal year 2013 as a range. Since data, such as laboratories names and addresses, as well as information on studies contained in OECA's LISA database are transferred from OPP's OPPIN, we also requested OPP provide information from OPPIN on the number of studies associated with performing GLP laboratories from fiscal year 2008 to fiscal year 2012. We analyzed this study data for accuracy and completeness. However, we determined that data for "performing laboratories," identification numbers, and addresses were not sufficiently reliable to assess the number and location of laboratories that had submitted studies to EPA. As a result, we were not able to define the universe of laboratories subject to a GLP inspection. However, EPA did provide estimates of the number and percentage of eligible laboratories that were inspected each year, and we used that information in this report.

To determine the views on EPA's GLP Compliance Monitoring Program, we interviewed individuals representing 25 entities, including laboratories, pesticide manufacturers, international organizations, environmental and health organizations, national and trade associations and foreign government GLP programs, which we selected based on referrals from various stakeholders and EPA officials. Specifically, we spoke with representatives of nine laboratories and conducted site visits to four of these laboratories located in Maryland and Illinois. We also conducted a Web-based survey of performing laboratories. For our survey, we drew a randomly selected sample of 80 performing laboratories included in the OPPIN database that sent study data to OPP from fiscal year 2010 to fiscal year 2012. Because OPPIN does not contain e-mail addresses, we searched for laboratories' e-mail addresses and found them for 53 of the 80 laboratories that we sampled. We sent the Web-based survey to these 53 laboratories. We obtained responses from 26 laboratories and other entities. However, 6 of these 26 respondents stated that they were not a laboratory and did not conduct GLP testing. Therefore, we received usable survey responses from 20 laboratories. Not all respondents to our survey, however, answered every question. Our survey asked if the laboratory had conducted GLP testing for EPA or FDA, experienced a GLP inspection by EPA, what was the effect of having an inspection, and what improvements they thought could be made to EPA's GLP Inspection Monitoring Program. The survey results are not generalizeable to all laboratories that are covered by the GLP program, but they were randomly selected, and the results can provide examples of such laboratories' experiences and challenges with GLP inspections. We were not able to select a representative sample because some data in the OPPIN database was not sufficiently reliable.

To assess the extent to which OPP uses the information obtained through GLP laboratory inspections in its pesticide decision-making process, we analyzed relevant documents and databases to determine the number of OECA's GLP inspections that produced results that were referred to OPP for reexamination, the time it took to conduct these reexaminations, and any impact the inspections had on OPP's pesticide registration decisions.

To determine the extent to which EPA and FDA collaborate on inspections, we analyzed EPA and FDA GLP laboratory inspection data and determined that the inspection data were sufficiently reliable for our purposes. We also analyzed the 20 laboratory survey responses to determine if they had conducted GLP tests or studies for submission to both EPA and FDA since 2008. We reviewed agency documents, such as a 1984 agreement entered into by EPA and FDA to cooperate on GLP inspections and FDA's Compliance Program Guidance Manual. We interviewed EPA and FDA officials, laboratory representatives and other stakeholders about the potential for the two agencies to collaborate.

We conducted this performance audit from November 2012 to May 2014 in accordance with generally accepted government auditing standards. Those standards require that we plan and perform the audit to obtain sufficient, appropriate evidence to provide a reasonable basis for our findings and conclusions based on our audit objectives. We believe that the evidence obtained provides a reasonable basis for our findings and conclusions based on our audit objectives.

Appendix II: Web-Based Survey to Laboratories

GAO Survey on EPA's Good Laboratory Practices Monitoring Program

United States Government Accountability Office

Background

As you may know, EPA relies on data submitted by pesticide registrants as the basis for the agency's regulatory decisions involving pesticide product registrations and tolerances. EPA and the Food and Drug Administration (FDA) have promulgated Good Laboratory Practices (GLP) regulations to assure the quality of data. EPA relies on data submitted by registrants as the basis for the agency's regulatory decisions involving pesticide product registrations and tolerances.

By "your facility," we are referring to the following lab: (Name of lab was pre-entered here)

Lab number: (Lab number was pre-entered here)

which we understand to be located at:
(Lab address was pre-entered here)

1. **Since January 2008, has your facility conducted testing or studies in accordance with GLP?**

○ Yes
○ No
○ Don't Know

If you selected "Don't know," please explain.

2. **Since January 2008, has EPA performed a GLP inspection of your facility?**

○ Yes
○ No
○ Don't Know

2a. **If YES, what effect, if any, did the GLP inspection have on your business?**

- ○ Mostly positive effect
- ○ Mostly negative effect
- ○ Neutral effect
- ○ No effect

What were these effects?

2b. **If NO, what effect, if any, did <u>not</u> having a GLP inspection have on your business?**

- ○ Mostly positive effect
- ○ Mostly negative effect
- ○ Neutral effect
- ○ No effect

2c. **What were these effects?**

If you selected "Don't know," please explain.

3. **In what ways, if at all, do you think EPA could improve its implementation of its GLP Compliance Monitoring Program?**

4. **Since 2008, were the GLP tests or studies your facility conducted intended for review by EPA, FDA or both agencies?**

- ○ EPA
- ○ FDA
- ○ Both agencies
- ○ Don't Know

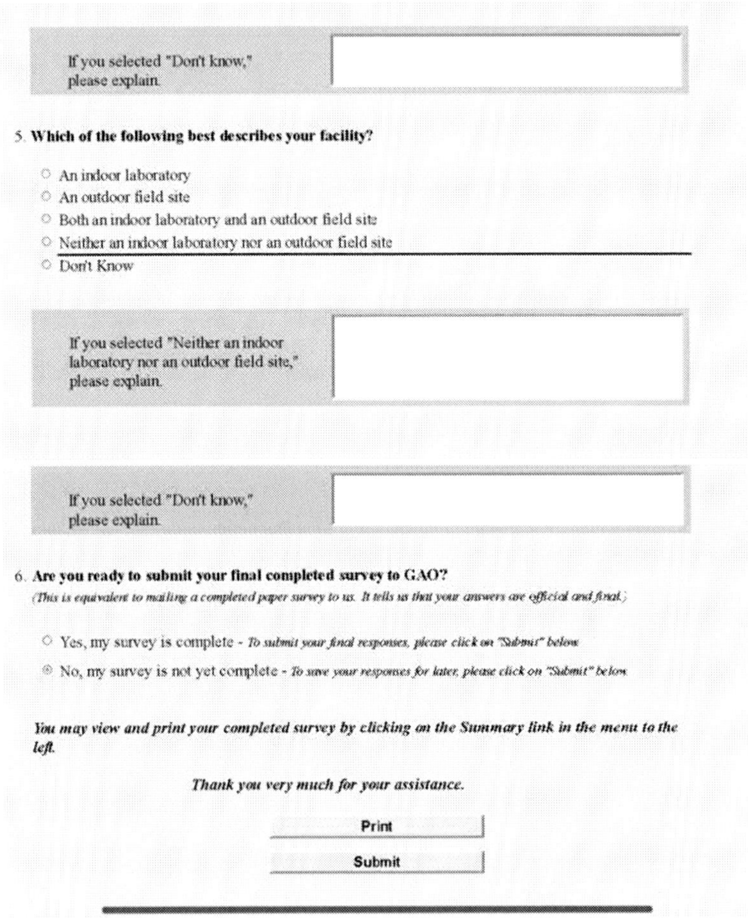
If you selected "Don't know," please explain.

5. **Which of the following best describes your facility?**

- An indoor laboratory
- An outdoor field site
- Both an indoor laboratory and an outdoor field site
- Neither an indoor laboratory nor an outdoor field site
- Don't Know

If you selected "Neither an indoor laboratory nor an outdoor field site," please explain.

If you selected "Don't know," please explain.

6. **Are you ready to submit your final completed survey to GAO?**
(This is equivalent to mailing a completed paper survey to us. It tells us that your answers are official and final.)

- Yes, my survey is complete - *To submit your final responses, please click on "Submit" below*
- No, my survey is not yet complete - *To save your responses for later, please click on "Submit" below*

You may view and print your completed survey by clicking on the Summary link in the menu to the left.

Thank you very much for your assistance.

Print

Submit

APPENDIX III: LABORATORIES INSPECTED FOR GLP COMPLIANCE BY EPA AND FDA, FY 2005-2012

Our analysis of Environmental Protection Agency (EPA) and U.S. Food and Drug Administration (FDA) Good Laboratory Practices (GLP) inspections data identified the following instances where laboratories were inspected by both agencies (see figure 3). We do not know the reason for every inspection. In some instances, EPA and FDA may have needed to conduct their own inspection.

Laboratory location	Fiscal year 2005	Fiscal year 2006	Fiscal year 2007	Fiscal year 2008	Fiscal year 2009	Fiscal year 2010	Fiscal year 2011	Fiscal year 2012
1. Laboratory in Alabama	FDA Sept. 2005	EPA Jan. 2006	FDA Feb. 2007		FDA Aug. 2009			FDA Sept. 2012
2. Laboratory in Arkansas		FDA Aug. 2006			FDA Mar. 2009	EPA May 2010		
3. Laboratory in California	FDA June 2005		EPA Aug. 2007 FDA Mar. 2007		FDA June 2009	EPA Feb. 2010		
4. Laboratory in Colorado	EPA Apr. 2005		FDA Mar. 2007					
5. Laboratory in Colorado	FDA July 2005	EPA Feb. 2006 FDA July 2006			EPA Dec. 2008 FDA Sept. 2009			
6. Laboratory in Colorado								EPA Nov. 2011 FDA June 2012
7. Laboratory in Georgia							EPA Sept. 2011	FDA June 2012

Figure 3. (Continued).

Laboratory location	Fiscal year 2005	Fiscal year 2006	Fiscal year 2007	Fiscal year 2008	Fiscal year 2009	Fiscal year 2010	Fiscal year 2011	Fiscal year 2012
8. Laboratory in Illinois	EPA Oct. 2004							EPA June 2012
		FDA Nov. 2005		FDA Mar. 2008, Sept. 2008			FDA Feb. 2011	
9. Laboratory in Illinois		EPA Nov. 2005		EPA Aug. 2008			EPA May 2011	
	FDA Jan. 2005		FDA Feb. 2007			FDA Sept. 2010		
10. Laboratory in Kansas	EPA Apr. 2005	EPA Sept. 2006		EPA Mar. 2008, Apr. 2008				
	FDA Apr. 2005	FDA Sept. 2006						
11. Laboratory in Kansas						EPA Sept. 2010		
							FDA June 2011	
12. Laboratory in Maryland	EPA July 2005			EPA June 2008				
								FDA Sept. 2012
13. Laboratory in Maryland			EPA Feb. 2007			EPA July 2010		EPA Sept. 2012
	FDA Sept. 2005		FDA inspected this lab in Sept. 2007 on behalf of EPA.	FDA Sept. 2008		FDA Aug. 2010	FDA Jan. 2011	
14. Laboratory in Massachusetts		EPA June 2006						
	FDA Mar. 2005			FDA Dec. 2007	FDA Apr. 2009		FDA Sept. 2011	

Figure 3. (Continued).

Laboratory location		Fiscal year 2005	Fiscal year 2006	Fiscal year 2007	Fiscal year 2008	Fiscal year 2009	Fiscal year 2010	Fiscal year 2011	Fiscal year 2012
15. Laboratory in Massachusetts	EPA				Aug. 2008				
	FDA	Sept. 2005							
16. Laboratory in Michigan	EPA				May 2008			Dec. 2010	
	FDA	Aug. 2005	May 2006	Apr. 2007	Sept. 2008	Nov. 2008	June 2010		Aug. 2012
17. Laboratory in Missouri	EPA		June 2006			May 2009			
	FDA		June 2006						July 2012
18. Laboratory in Missouri	EPA							Aug. 2011	
	FDA		Nov. 2005		June 2008; Aug. 2008				
19. Laboratory in Montana	EPA					Sept. 2009			
	FDA		Jan. 2006						
20. Laboratory in New Jersey	EPA	Nov. 2004		Feb. 2007		Jan. 2009			
	FDA		Apr. 2006			Sept. 2009		July 2011	
21. Laboratory in New Jersey	EPA						Dec. 2009		
	FDA	Jan. 2005	Aug. 2006			Dec. 2008			

Figure 3. (Continued).

Laboratory location	Fiscal year 2005	Fiscal year 2006	Fiscal year 2007	Fiscal year 2008	Fiscal year 2009	Fiscal year 2010	Fiscal year 2011	Fiscal year 2012
22. Laboratory in New Mexico			EPA Aug. 2007			EPA Mar. 2010		
	FDA Feb. 2005				FDA Aug. 2009			
23. Laboratory in New Mexico	EPA May 2005							
			FDA Dec. 2006		FDA Aug. 2009			
24. Laboratory in North Carolina		EPA Mar. 2006						
	FDA Nov. 2004		FDA Sept. 2007					FDA Aug. 2012
25. Laboratory in North Carolina				EPA Aug. 2008		EPA Sept. 2010		
	FDA Jan. 2005		FDA June 2007					
26. Laboratory in Ohio					EPA Apr. 2009			
	FDA Mar. 2005		FDA Aug. 2007			FDA Oct. 2009	FDA Jan. 2011	
27. Laboratory in Ohio		EPA Sept. 2006			EPA Apr. 2009			
		FDA Feb. 2006		FDA Sept. 2008	FDA Aug. 2009; Aug. 2009	FDA Jan. 2010		FDA Apr. 2012
28. Laboratory in Ohio			EPA Aug. 2007					
			FDA Mar. 2007		FDA Sept. 2009			

Figure 3. (Continued).

Laboratory location		Fiscal year 2005	Fiscal year 2006	Fiscal year 2007	Fiscal year 2008	Fiscal year 2009	Fiscal year 2010	Fiscal year 2011	Fiscal year 2012
29. Laboratory in Ohio	EPA					May 2009			Dec. 2011
	FDA			Oct. 2006		July 2009			
30. Laboratory in Pennsylvania	EPA			June 2007		Feb. 2009			June 2012
	FDA	Dec. 2004			Oct. 2007		May 2010		
31. Laboratory in Pennsylvania	EPA	Aug. 2005			Apr. 2008				
	FDA			Oct. 2006		May 2009	June 2010		
32. Laboratory in Pennsylvania	EPA			Nov. 2006					
	FDA		Feb. 2006			June 2009	Aug. 2010		July 2012
33. Laboratory in Texas	EPA		Dec. 2005				Apr. 2010		Sept. 2012
	FDA	Sept. 2005		Jan. 2007			Apr. 2010		
34. Laboratory in Utah	EPA								Jan. 2012
	FDA	Aug. 2005							June 2012
35. Laboratory in Virginia	EPA		Aug. 2006			Oct. 2008			
	FDA	May 2005		July 2007	May 2008	July 2009; July 2009		Aug. 2011	

Figure 3. (Continued).

Laboratory location	Fiscal year 2005	Fiscal year 2006	Fiscal year 2007	Fiscal year 2008	Fiscal year 2009	Fiscal year 2010	Fiscal year 2011	Fiscal year 2012
36. Laboratory in Wisconsin	EPA Aug. 2005							EPA Sept. 2012
	FDA Jan. 2005		FDA May 2007	FDA June 2008		FDA Aug. 2010		
37. Laboratory in Wisconsin			EPA June 2007				EPA Aug. 2011	
		FDA July 2006		FDA Sept. 2008			FDA May 2011	

EPA performed inspection FDA performed inspection

Source: GAO analysis of EPA and FDA data.

Figure 3. Laboratories Inspected for GLP Compliance by EPA and FDA, FY 2005-2012 .

End Notes

[1] Act of June 25, 1947, ch. 125, 61 Stat. 163 (codified as amended at 7 U.S.C. §§ 136-136y).
[2] 7 U.S.C. § 136a(a).
[3] 40 C.F.R. pts. 152-180.
[4] EPA implements two sets of regulations setting forth GLP standards—one under FIFRA, see 40 C.F.R. pt. 160, and one under the Toxic Substances Control Act (TSCA), Pub. L. No. 94-469, 90 Stat. 2003 (1976) (codified as amended at 15 U.S.C. §§ 2601-2697), see 40 C.F.R. pt. 792. According to EPA officials, most of its GLP inspections are conducted under FIFRA. For that reason, we have focused on GLP standards under FIFRA in this report.
[5] A "laboratory" in this report is defined as a facility where a study or part of a study is conducted. A laboratory can conduct its testing either indoors or outdoors at a field site.
[6] EPA regulations define a sponsor as (1) a person who initiates and supports, by the provision of financial or other resources, a study; (2) a person who submits a study to EPA in support of an application for a research or marketing permit; or (3) a testing facility, if it both initiates and actually conducts the study. 40 C.F.R. § 160.3.
[7] We are using the term "audit" to refer to the OECA inspectors' examination of studies.
[8] Act of June 25, 1938, ch. 675, 52 Stat.1040 (codified as amended at 21 U.S.C. §§ 301-399f).
[9] 21 C.F.R. pt. 58.
[10] EPA reviews health and environmental effects data submitted by a company and may register a pesticide or, alternatively, grant a "conditional registration" for a pesticide under certain circumstances, even though some of the required data may not have been submitted or reviewed. We recommended, in part, that EPA consider and implement options for an automated system to better track conditional registrations. See GAO, *Pesticides: EPA Should Take Steps to Improve Its Oversight of Conditional Registrations*, GAO-13-145 (Washington, D.C.: Aug. 8, 2013).
[11] While we received a total of 26 survey responses from our sample of laboratories, 6 of the 26 reported that they were not a laboratory and did not conduct GLP testing. Therefore, these 6 are considered out-of-scope for reporting our survey results. See appendix I for more detail.
[12] The phrase "unreasonable adverse effects on the environment" is defined in FIFRA, in part, to mean (1) any unreasonable risk to man or the environment, taking into account the economic, social, and environmental costs and benefits of the use of any pesticide or (2) a human dietary risk from residues that result from a use of a pesticide in or on any food inconsistent with the standard for tolerance under section 408 of the Federal Food, Drug, and Cosmetic Act, 21 U.S.C. § 346a. This standard requires EPA to consider both the benefits and risks of using a pesticide. 7 U.S.C. § 136(bb).
[13] Under section 408 of the FFDCA, "safe" means that EPA has determined, among other things, that there is a reasonable certainty that no harm will result from aggregate exposure to the pesticide residue. 21 U.S.C. § 346a(b)(2)(A)(ii).
[14] According to OPP officials, applicants for pesticide registrations are usually pesticide product manufacturers. When EPA registers an applicant's pesticide product, the applicant is then called a registrant. For the purposes of this report, we refer to both applicants and registrants as registrants.
[15] 40 C.F.R. § 160.12.
[16] 43 Fed. Reg. 59,986 (Dec. 22, 1978) (codified as amended at 21 C.F.R. pt. 58).
[17] OECD is an organization of 34 industrialized countries, operating by consensus, that fosters dialogue among members to discuss, develop, and refine economic and social policies and provides an arena for establishing multilateral agreements.
[18] EPA has signed a memorandum of understanding with Canada, Germany, Israel, Japan, the Netherlands, Switzerland, and the United Kingdom. In addition, EPA has signed a Letter of Intent with China to mutually accept each other's GLP Program upon completion of certain criteria, and the U.S. Ambassador signed a bilateral agreement with Taiwan.

[19] When pesticide registration application or tolerance level petitions are submitted to OPP, contractors enter laboratory data, such as laboratory name and address, as well as information on the studies submitted with the application, into the Office of Pesticide Program Information Network (OPPIN) database. At any given time, LISA only contains data submitted to OPP for the previous 5 years. The data are transferred to LISA from OPPIN. In contrast to EPA's process, officials from other countries told us that their GLP compliance monitoring programs select laboratories for inspection before these laboratories begin GLP testing.

[20] Some laboratories are identified for inspection through other means, such as complaints submitted.

[21] According to EPA officials, approximately 1,400 laboratories conducted studies that were submitted to EPA over a 5-year time frame and are thus eligible for inspection.

[22] FDA Compliance Program Guidance Manual 7348.808.

[23] Not all respondents to our survey answered every question. We obtained survey responses from 20 laboratories from a nongeneralizable, random sample of 53 laboratories, and other entities that had sent supporting studies to OPP for a pesticide registration application for fiscal year 2010 to 2012 to ask them questions about EPA's GLP inspections. Please see appendix I for details.

[24] We have previously reported issues with inaccurate data in OPPIN. In August 2013, we reported that, according to OPP officials, following an internal review of its conditional pesticide registration program that it completed in March 2011, OPP concluded that the OPPIN data on the number of conditional pesticide registrations were inaccurate. See GAO-13-145.

[25] In the *Pesticide Registration (PR) Notice 2011-3 Standard Format for Data Submitted Under the Federal Insecticide, Fungicide, and Rodenticide Act (FIFRA) and Certain Provisions of the Federal Food, Drug, and Cosmetic Act (FFDCA)* published on November 30, 2011, EPA instructs study submitters to clearly identify on each study's title page the name of the performing laboratory. The guidance does not specify that a laboratory's identification number be included on the title page but does specify inclusion of a laboratory project identifier.

[26] We received a total of 26 survey responses from our sample of laboratories, but 6 of the 26 reported that they were not a laboratory and did not conduct GLP testing. Therefore, these 6 are considered out-of-scope for reporting our survey results.

[27] Missing address information could include missing data for any of the following: street address, city, state, or zip code.

[28] EPA, *Technical Guidance for Citation Building: Extra Hints for Coding Laboratories and Sponsors*. Apr. 1, 2008.

[29] EPA, *Revised Technical Direction for PDMS Laboratory Indexing*. Nov. 1991.

[30] GAO, *2012 Annual Report: Opportunities to Reduce Duplication, Overlap and Fragmentation, Achieve Savings, and Enhance Revenue*, GAO-12-342SP (Washington, D.C.: February 2012).

[31] GAO, *Federal User Fees: Fee Design Options and Implications for Managing Revenue Instability*, GAO-13-820 (Washington, D.C.: Sept. 30, 2013).

[32] In May 2008, we issued a User Fee Design Guide, which examined how the four key design and implementation characteristics—how fees are set, collected, used, and reviewed—may affect the economic efficiency, equity, revenue adequacy, and administrative burden of the fees. GAO, *Federal User Fees: A Design Guide*, GAO-08-386SP (Washington, D.C.: May 2008).

[33] Office of Mgmt. & Budget, Circular No. A-25 Revised, Memorandum for Heads of Executive Departments and Establishments (July 8, 1993).

[34] The rule governing the accounting and disposition of receipts, including user fees, states that they must be deposited in the general fund of the Treasury as miscellaneous receipts unless the agency has statutory authority to do otherwise. 31 U.S.C. § 3302.

[35] Pub. L. No. 101-576, 104 Stat. 2838 (1990).

[36] EPA, *Fiscal Year 2013 Agency Financial Report.*

[37] EPA Office of Inspector General, *EPA Did Not Conduct Thorough Biennial User Fee Reviews*, 14-P-0129 (Washington, D.C.: Mar. 4, 2014).

[38] EPA acknowledges that improvements were needed to ensure that all agency programs were reviewed, in accordance with OMB Circular A-25, "User Charges, Revised," and the Chief Financial Officers Act of 1990. According to EPA officials, the agency took actions in Fiscal Year 2013 to improve its user fee review efforts by issuing a biennial user fee review guide, conducting training for its user fee programs, and established fee review procedures.

[39] GAO-08-386SP.

[40] The Pesticide Registration Improvement Act of 2003 (PRIA) provides a schedule of covered applications and registration service fees, including the category or type of application, the amount of the pesticide registration service fee, and the corresponding decision review timeframe in which the agency is to make a decision. There are 189 fee categories or types of applications, each with a fee and decision review time frame. Pub. L. No. 108-199, div. G, tit. V, § 501(f)(2), 118 Stat. 419, 422 (2004) (codified as amended at 7 U.S.C. § 136w-8).

[41] GAO, *Standards for Internal Control in the Federal Government*, GAO/AIMD-00-21.3.1 (Washington, D.C.: November 1999).

[42] We have defined duplication as occurring when two or more agencies or programs engaged in the same activities or provided the same services to the same beneficiaries.

[43] FDA continued to perform inspections for EPA until 2007, although the agreement was not renewed after 2004.

[44] FDA consists of nine centers and offices. According to FDA officials, five of these centers conduct GLP inspections: the Center for Biologics Evaluation and Research, the Center for Devices and Radiological Health, the Center for Drug Evaluation and Research, the Center for Food Safety and Applied Nutrition, and the Center for Veterinary Medicine.

[45] EPA officials provided us with an analysis that identified GLP inspections conducted by EPA and FDA at 31 laboratories. We compared these 31 laboratories with our analysis that identified 37 laboratories. We found that EPA's analysis included some additional laboratories that we could not find in our analysis. Conversely, EPA's analysis excluded some laboratories that were included in our analysis, which was based on a review of inspection data provided to us by EPA and FDA.

[46] We have defined overlap as two or more agencies or programs engaging in similar activities or providing similar services to similar beneficiaries.

[47] GAO, *2013 Annual Report: Actions Needed to Reduce Fragmentation, Overlap, and Duplication and Achieve Financial Benefits*, GAO-13-279SP (Washington, D.C.: April 2013).

[48] GAO, *Managing for Results: Key Considerations for Implementing Interagency Collaborative Mechanisms,* GAO-12-1022 (Washington, D.C.: Sept. 27, 2012).

[49] GAO, *Results-Oriented Government: Practices That Can Help Enhance and Sustain Collaboration among Federal Agencies*, GAO-06-15 (Washington, D.C.: Oct. 21, 2005).

[50] GAO-13-279SP.

[51] These statutes and regulations include Federal Insecticide, Fungicide, and Rodenticide Act, Act of June 25, 1947, ch. 125, 61 Stat. 163 (codified as amended at 7 U.S.C. §§ 136-136y); Federal, Food, Drug, and Cosmetic Act, Act of June 25, 1938, ch. 675, 52 Stat. 1040 (codified as amended at 21 U.S.C. §§ 301-399f); 21 C.F.R. pt. 58; 40 C.F.R. pts. 152-180.

[52] These program and guidance documents included *EPA, Good Laboratory Practice Standards, Inspection Manual*, (Washington, D.C.: Sept. 1993); *Enforcement Response Policy for the Federal Insecticide, Fungicide, and Rodenticide Act, Good Laboratory Practices (GLP) Regulations, Sept. 1991.*

[53] GAO/AIMD-00-21.3.1.

[54] Office of Mgmt. & Budget, Circular No. A-25 Revised, Memorandum for Heads of Executive Departments and Establishments (July 8, 1993).

[55] GAO-06-15; EPA Office of Inspector General, *EPA Did Not Conduct Thorough Biennial User Fee Reviews*, 14-P-0129 (Washington, D.C.: Mar. 4, 2014).

In: EPA Oversight of Pesticides
Editor: Maggie Gibson

ISBN: 978-1-63463-177-8

Chapter 2

PESTICIDES: EPA SHOULD TAKE STEPS TO IMPROVE ITS OVERSIGHT OF CONDITIONAL REGISTRATIONS*

United States Government Accountability Office

WHY GAO DID THIS STUDY

As of September 2010, more than 16,000 pesticides were registered for use in the United States, according to EPA. EPA reviews health and environmental effects data submitted by a company and may register a pesticide or, alternatively, grant a "conditional registration" for a pesticide under certain circumstances, even though some of the required data may not have been submitted or reviewed. The company must provide the missing data within a specified time. In 2010, environmental and other groups charged that EPA had overused conditional registrations and did not appear to have a reliable system to identify whether the required data had been submitted. GAO was asked to examine issues related to EPA's use of conditional registrations for pesticides. This report examines the (1) number of conditional registrations EPA has granted and the basis for these, (2) extent to which EPA ensures that companies submit the required additional data and EPA reviews the data, and (3) views of relevant stakeholders on EPA's use of

* This is an edited, reformatted and augmented version of the United States Government Accountability Office publication, GAO-13-145, dated August 2013.

conditional registrations. GAO reviewed EPA data and surveyed stakeholders, among other things.

What GAO Recommends

GAO recommends, in part, that EPA consider and implement options for an automated system to better track conditional registrations. EPA agreed with GAO's recommendations and noted specific actions it will take to implement them.

What GAO Found

The total number of conditional registrations granted is unclear, as the Environmental Protection Agency (EPA) reports that its data are inaccurate for several reasons. First, the database used to track conditional registrations does not allow officials to change a pesticide's registration status from conditional to unconditional once the registrant has satisfied all requirements, thereby overstating the number of conditional registrations. Second, EPA staff have misused the term "conditional registration," incorrectly classifying pesticide registrations as conditional when, for example, they require a label change, which is not a basis in statute for a conditional registration. According to EPA documents and officials, weaknesses in guidance and training, management oversight, and data management contributed to these misclassification problems. For example, according to EPA documents, there was limited, organized management oversight to ensure that regulatory actions were not misclassified as conditional registrations. As of July 2013, EPA officials told GAO that the agency has taken or is planning to take several actions to more accurately account for conditional registrations, including beginning to design a new automated data system to more accurately track conditional registrations.

The extent to which EPA ensures that companies submit additional required data and EPA reviews these data is unknown. Specifically, EPA does not have a reliable system, such as an automated data system, to track key information related to conditional registrations, including whether companies have submitted additional data within required time frames. As a result, pesticides with conditional registrations could be marketed for years without EPA's receipt and review of these data. In the absence of a reliable system for managing conditional registrations, EPA relies on a variety of routine program operations, such as its review of a company's changes to a pesticide registration, to discover that data are missing. However, these methods fall short of what is needed because they are neither comprehensive nor do they ensure timely submission of these data. According to federal internal control standards, EPA's lack of a reliable system for managing conditional registrations constitutes an internal control weakness because the agency lacks an effective mechanism for program oversight and decision making.

Stakeholders GAO surveyed—representatives of consumer, environmental, industry, legal, producer, science, and state government groups—generally said EPA needs to improve its conditional registration process. For example, some stated EPA should improve its data systems for tracking conditional registrations to ensure that required data are submitted and reviewed in a timely manner. However, stakeholder views varied on the benefits and disadvantages of conditionally registering pesticides. For example, some consumer, industry, legal, producer, and state government stakeholders stated that the conditional registration process promotes innovation by bringing new technologies to the marketplace more quickly. In contrast, some consumer, environmental, legal, science, and state government stakeholders voiced concerns that conditional registration allows products with safety that has not been fully evaluated into the marketplace.

ABBREVIATIONS

DCI	Data Call-In
EPA	Environmental Protection Agency
FIFRA	Federal Insecticide, Fungicide, and Rodenticide Act
FFDCA	Federal Food, Drug, and Cosmetic Act
NRDC	Natural Resources Defense Council
OMB	Office of Management and Budget
OPP	Office of Pesticide Programs
OPPIN	Office of Pesticide Programs Information Network
PRIA	Pesticide Registration Improvement Act of 2003
PRISM	Pesticide Registration Information System

August 8, 2013
The Honorable Paul D. Tonko
Ranking Member
Subcommittee on Environment and the Economy
Committee on Energy and Commerce
House of Representatives

Dear Mr. Tonko:

As of September 2010, more than 16,000 pesticides were registered for use in the United States, according to the Environmental Protection Agency (EPA). In addition, expenditures for pesticides—primarily for use in agriculture—totaled $12.5 billion in 2007, the latest year for which data are available from EPA. These pesticides—chemicals or biological substances used to destroy or control weeds or unwanted insects, fungi, rodents, bacteria and other pests—contribute significantly to agricultural productivity by preventing crop damage and improving public health by controlling disease-carrying pests. However, if used improperly, pesticides may adversely affect human health by, for example, increasing the risk of cancer and neurological disorders caused by pesticide residues on food crops. Pesticides may also damage the environment by, for example, killing or causing reproductive abnormalities in species other than those which the pesticide is intended to kill, including fish, birds, and other wildlife.

Under the Federal Insecticide, Fungicide, and Rodenticide Act, as amended, (FIFRA),[1] EPA registers pesticides distributed, sold, or used in the United States and prescribes labeling and other regulatory requirements to prevent unreasonable adverse effects on the environment.[2] To obtain a registration, a company or person (registrant) is to submit an application containing health and environmental effects data and other information on a pesticide for EPA's review.[3] After reviewing the information, in general, EPA may (1) register the pesticide and set a tolerance level (i.e., the maximum pesticide residue allowed) for those pesticides used on food or animal feed,[4] (2) notify the registrant of deficiencies in the data or the need for additional information, or (3) deny the application.

FIFRA section 3(c)(5) and section 3(c)(7) set forth two sets of statutory standards under which EPA may register pesticides. EPA is authorized to grant registrations under section 3(c)(5) if EPA determines, among other things, that the application materials provided satisfy the requirements of the statutes and that the information those materials contain concerning the pesticide demonstrates that, when used in accordance with widespread and commonly recognized practice, the product will not generally cause unreasonable adverse effects on the environment. Such registrations are commonly referred to as "unconditional" registrations because they are approved without EPA imposing a requirement on the registrant to develop additional data.

In addition, section 3(c)(7) of FIFRA provides EPA the authority to grant a "conditional registration" for a pesticide product under certain circumstances, although some necessary data have not been provided by the registrant in the application. Under this authority, EPA may grant a conditional registration for pesticide products that are identical or substantially similar to pesticide products that are already registered or for a new use of a currently registered product as long as EPA determines, among other things, that the pesticide will not significantly increase the risk of unreasonable adverse effects on the environment. This authority also allows EPA to grant a conditional registration for pesticide products with new active ingredients where data are missing from the application because the requirement for the data was imposed so recently that the registrant did not have adequate time to generate the data and EPA determines, among other things, that the pesticide will not cause unreasonable adverse effects on the environment during the time needed to generate the data.[5] According to EPA officials, registrants typically have from 1 to 4 years to provide the missing data required by a conditional registration.

In 2010, an environmental group questioned EPA's use of conditional registrations in the pesticide program, and other consumer and science groups supported this position. The environmental group charged that EPA had overused conditional registrations, stating that information from EPA's Office of Pesticide Programs Information Network (OPPIN) data system showed that conditional registrations represented the majority of active registrations.[6] The group also pointed out that the OPPIN data indicated that some pesticide products had been conditionally registered for 20 years or more.

In this context, you asked us to examine EPA's procedures for granting conditional registrations. Accordingly, we examined the (1) number of conditional pesticide registrations EPA has granted and the basis for granting these registrations; (2) extent to which EPA ensures that registrants submit the additional data required as part of conditional registrations and reviews these data; and (3) views of relevant stakeholders on EPA's use of conditional registrations, including ways, if any, to improve the conditional registration process.

To address these objectives, we reviewed relevant federal statutes and regulations, EPA guidance documents and internal reviews, federal internal control standards, and previous GAO and EPA Inspector General reports.[7] We also requested that EPA's Office of Pesticide Programs (OPP) provide summary data on its issuance of conditional registrations, including (1) the number of pesticide registrations currently in conditional status and how long they have been in this status; (2) the total number of current pesticide registrations (conditional and unconditional); and (3) for fiscal years 1997 through 2011, the number of conditional registrations granted for each year and the basis on which these were granted (i.e., identical/substantially similar pesticides, new uses, or new active ingredients). In addition, we asked OPP to provide information on the number of pesticides that (1) have a conditional registration, but the registrant has not submitted the additional required data by the specified due date; (2) have a conditional registration and the registrant has submitted the additional required data, but EPA has not reviewed these data; (3) still have a conditional registration, even though the registrant has submitted, and EPA has reviewed, the additional required data; and (4) had been changed from conditional to unconditional status. However, after interviewing OPP officials and reviewing past GAO, Inspector General, and EPA contractor studies examining OPP's data management, especially its use of OPPIN, we concluded that EPA could not provide us with sufficiently reliable data for obtaining summary level information on conditional pesticide registrations. In the absence of these data, we discussed with OPP officials the data limitations they face using OPPIN and any potential workarounds they employ or are planning. In addition, we interviewed OPP officials and reviewed documentation they provided to obtain further information and clarification on EPA's conditional registration process, including any planned responses to internal review or external stakeholder concerns. To obtain the views of relevant stakeholders on EPA's conditional registration process and ways, if any, to improve it, we used the results of a literature search and other sources to develop a list of consumer, environmental, industry, legal, producer, science, and state government stakeholders. From this list, we selected a nonprobability sample of stakeholders from each category to contact, for a total of 35 stakeholders.[8] We then developed, pretested, and e-mailed a questionnaire to obtain these stakeholders' views. Twenty-four out of 35 stakeholders provided responses to our questionnaire. We performed a content analysis of the questionnaire responses to identify common themes regarding stakeholders' views on EPA's use of conditional registrations. Appendix I provides a more detailed description of our objectives, scope, and methodology.

We conducted this performance audit from September 2011 to August 2013 in accordance with generally accepted government auditing standards. Those standards require that we plan and perform the audit to obtain sufficient, appropriate evidence to provide a reasonable basis for our findings and conclusions based on our audit objectives. We believe that the evidence obtained provides a reasonable basis for our findings and conclusions based on our audit objectives.

BACKGROUND

The primary federal laws that govern how EPA regulates pesticides in the United States are FIFRA and the Federal Food, Drug, and Cosmetic Act (FFDCA).[9] Under FIFRA implementing regulations,[10] EPA is to review applications for pesticide products and register those that it determines will meet the FIFRA statutory standards for registration. If the use of a pesticide would result in a residue of the substance in or on food or animal feed, EPA may not register a pesticide under FIFRA unless it can determine that the residue is "safe" as defined by FFDCA. Under FFDCA, safe means that EPA has determined, among other things, that there is a reasonable certainty that no harm will result from aggregate exposure to the pesticide residue, including all anticipated dietary exposures and all other nonoccupational exposures for which there is reliable information.[11] EPA may establish a tolerance level—the maximum permissible pesticide residue in or on food or animal feed that is sold—that meets the FFDCA safety standard or may choose to grant an exemption for a tolerance.[12]

OPP—the EPA office primarily responsible for regulating the use of pesticides—has regulatory staff in three divisions—Registration, Biopesticides and Pollution Prevention, and Antimicrobials—that are responsible for registering pesticides.[13] The registration process formally begins when a registrant submits an application to OPP for a particular pesticide. This application is to include data to support the registration of the pesticide. In reviewing the application, OPP is to examine, among other things, the pesticide's ingredients; the site or crop on which it is to be used; the amount, frequency, and timing of its use; and storage and disposal practices. OPP is also to review toxicity tests and studies showing how the pesticide affects human health and the environment. According to OPP officials, the length of time OPP takes from the initial review of an application to the final decision on whether to register a pesticide depends on many factors—including whether the pesticide being reviewed is similar to any pesticide EPA has previously reviewed— and, according to OPP officials, can take from 3 to 24 months. After OPP completes its review and approves the submitted package, EPA may register the pesticide without imposing requirements for additional data (unconditional registration) under FIFRA 3(c)(5) if EPA determines, among other things, that use of the pesticide in accordance with label directions will not have unreasonable adverse effects on the environment.[14]

Alternatively, FIFRA section 3(c)(7) allows EPA to grant a conditional registration for pesticides in the following circumstances:

- *Identical/substantially similar pesticides (FIFRA Section 3(c)(7)(A)).* EPA may conditionally approve an application for registration or an amended registration for a pesticide product if the agency determines that
    - the pesticide and proposed use are identical or substantially similar to any currently registered pesticide and its uses, or differ only in ways that will not significantly increase the risk of unreasonable adverse effects on the environment; and
    - approving the registration or amendment in the manner proposed would not significantly increase the risk of any unreasonable adverse effect on the environment.

Each registration issued under 3(c)(7)(A) must submit or cite the same data that would be required for the unconditional registration of a similar product.

- *New uses (FIFRA Section 3(c)(7)(B)).* A current pesticide registration may be amended to allow additional uses, even if the data concerning the pesticide may be insufficient to support unconditional registration, if EPA determines that
    - the applicant has submitted satisfactory data pertaining to the proposed additional use; and
    - amending the registration would not significantly increase the risk of unreasonable adverse effect on the environment.

Each registrant must submit or cite the same data that would be required for the unconditional registration of a similar product.[15]

- *New active ingredients (FIFRA Section 3(c)(7)(C)).* A pesticide containing a new active ingredient not found in any currently registered pesticide can be conditionally registered for a period reasonably sufficient for the generation and submission of required data, if EPA determines
  - insufficient time has elapsed since the imposition of the data requirement for those data to be developed and on the condition that when the agency receives such data that they do not meet or exceed risk criteria stated in the regulations issued under FIFRA and other conditions issued by the agency;
  - the use of the pesticide during the period of the conditional registration will not cause unreasonable adverse effect on the environment; and
  - the use of the pesticide is in the public interest.

After a pesticide product is conditionally registered under FIFRA section (3)(c)(7), the registrant receives a notice indicating the terms of the conditional registration, including a list of any additional data that will need to be submitted and deadlines for submitting these data. Figure 1 summarizes the pesticide registration and tolerance setting process.

According to EPA officials, once a pesticide is conditionally registered, EPA typically grants a period of time, generally 1 to 4 years, for the registrant to provide the required data. The registrant can ask EPA to waive the requirement for additional information or, according to EPA officials, extend the time frame. If the registrant does not submit the data specified within the required time frame, EPA can cancel the pesticide registration.

Before a pesticide can be sold or distributed in the United States, it must be registered under FIFRA.[16] At any time, EPA may initiate a suspension or cancellation proceeding for a pesticide registration if safety concerns develop. For example, EPA began proceedings to cancel some uses of Carbofuran[17]—an insecticide and nematicide that was registered to control pests in soils and on leaves in a variety of field, fruit, and vegetable crops—after the agency determined that the dietary, worker, and ecological risks of this pesticide were unacceptable.[18]

Another check on the safety of registered pesticide products is the requirement in FIFRA section 6(a)(2) and FIFRA implementing regulations that registrants report adverse effects-related information to EPA. For example, registrants are required to submit certain toxicity information concerning the product both before and after registration, such as information on the product's toxicity to nontarget plant species.

In addition, as required by FIFRA,[19] EPA is to review the safety of each registered pesticide every 15 years to help ensure that each pesticide registration continues to satisfy the regulatory standard. In 2007, EPA began conducting these reviews under its registration review program.[20] As a part of this program, if EPA determines that additional data are needed to support the continued registration of a pesticide, the agency may issue a Data Call-In (DCI) notice, as authorized by FIFRA section 3(c)(2)(B), requiring the registrant to provide the data by a specific date. Also, at any time after a pesticide is registered, a registrant may apply to amend the registration and, according to OPP officials, such requests are reviewed as though the registrant is seeking approval for a new pesticide. While reviewing such requests, the agency may also issue a DCI notice requiring the registrant to provide additional data by a specific date. If a registrant fails to provide the data requested through a DCI, EPA may suspend the pesticide's registration under authority of FIFRA section 3(c)(2)(B).

In 2003 amendments to FIFRA,[21] pesticide registration fees paid by registrants were established for some registration actions, such as registrations for new uses of pesticides, to help pay for registration costs. Earlier amendments to FIFRA, enacted in 1988, established annual registration maintenance fees—fees used to support the review of existing pesticide registrations. For fiscal year 2012, EPA reported that it collected a total of \$37.6 million from fees, including \$22 million in maintenance fees.[22]

In July 2010, an environmental group requested information from EPA regarding conditional registrations, including the number of conditional registrations the agency had issued, whether registrants had submitted the additional data required by the registrations, and whether EPA had reviewed the data submitted.[23] OPP provided the group with information from the OPPIN data system on the number of pesticide registrations that had been categorized as conditional. Subsequently, in September 2010, the environmental group raised concerns about EPA's use of its conditional registration authority, and several other environmental groups and other interested parties supported this position. Among other things, the environmental group asserted that EPA had overused conditional registrations and did not appear to have a reliable tracking system to identify the status of conditionally registered pesticides to ensure that registrants submitted, and EPA reviewed, additional data in a timely manner.[24] In addition, the group noted that the information EPA provided indicated that many pesticides have remained in conditional status for many years. For example, the information showed that over 3,200 pesticides had been in conditional status since 1995 (15 years) and that 2,100 pesticides had been in conditional status since 1990 (20 years).

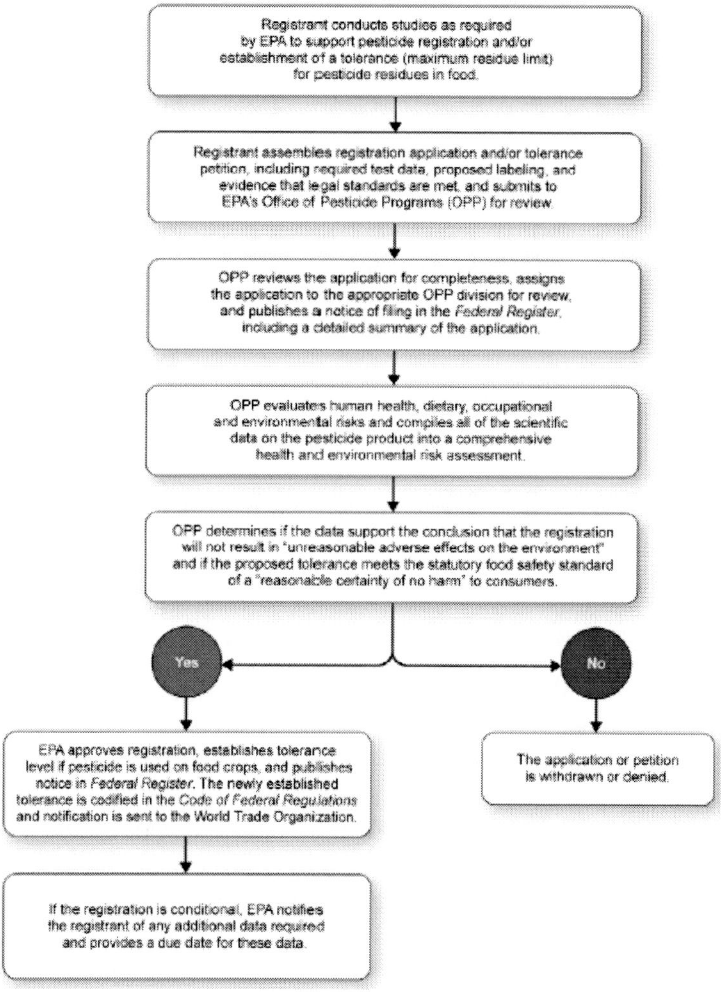

Source: GAO analysis of EPA documents.

Figure 1. General Overview of EPA Pesticide Registration and Tolerance Setting Process.

TOTAL NUMBER OF CONDITIONAL REGISTRATIONS GRANTED IS UNCLEAR, AS EPA FOUND DATA TO BE INACCURATE

The number of active conditional registrations EPA has granted is unclear, according to OPP officials who, as a result of a 2011 EPA review, found the agency's registration data to be inaccurate and the basis for granting some of these registrations to be inappropriately classified. Specifically, an internal review of OPP's conditional registration program found that OPPIN does not allow officials to change a pesticide's registration status from conditional to unconditional once the registrant has satisfied all data requirements, and the basis for many registration decisions was mischaracterized as conditional. In addition, based on the internal review, OPP officials concluded that several weaknesses contributed to this misclassification problem, including insufficient guidance and training, management oversight, and data management. As of July 2013, OPP officials told us that the office has taken or is planning to take several actions to more accurately account for conditional registrations.

EPA Found Conditional Registration Data to Be Inaccurate and That the Basis for Many Registrations Was Inappropriately Classified

According to OPP officials, following an internal review of its conditional registration program that it completed in March 2011, OPP concluded that the OPPIN data on the number of conditional registrations were inaccurate. The internal review was conducted, in part, to determine the number of conditional registrations granted by EPA. According to information on OPP's website,[25] during the internal review, OPP determined that OPPIN contained 16,156 active pesticide registrations and that 11,205 (69 percent) of these pesticides were conditionally registered. However, OPP officials concluded, based on the internal review, that the data were inaccurate, and that the number of conditional registrations was overstated, for two reasons. First, once a registration is classified as conditional in OPPIN, its status in this data system cannot be changed from conditional to unconditional, when, for example, the registrant has satisfied all data requirements imposed. According to the internal review, OPPIN is an older system that was not designed specifically to track conditional registrations and thus is ill-suited for that purpose. To determine the current number of conditionally registered pesticides, OPP officials said detailed paper files that support each pesticide registration would need to be reviewed, which would be a very time-consuming process. OPP officials indicated that they plan to develop a new automated system for tracking conditional registrations and, in fiscal year 2013, they began using a portion of the registration maintenance fees collected annually to begin exploring the feasibility of implementing such a system.[26] However, these officials were uncertain about the ultimate cost of this system, what sources they would use for any additional funding, and when the system would be operational.

Second, as a result of the internal review, OPP found that its regulatory staff had incorrectly categorized the basis for many program actions in OPPIN as "conditional registrations" and that these incorrect categorizations resulted in an overcounting of conditional registrations. According to the internal review, OPP staff had used conditional registrations to describe a variety of actions that fall outside of the circumstances authorized by FIFRA Section 3(c)(7). For example, according to OPP officials and the internal review, OPP staff had assigned the category "conditional registration" to situations where approval of a registration is contingent upon the "condition" that the registrant makes a change that does not involve generating additional data. These situations included certain changes to pesticide product labels—such as strengthening precautionary statements—that are not specified by FIFRA section 3(c)(7), according to the results of the internal review. Similarly, OPP staff categorized as "conditional registrations" situations where the agency requested registrants to provide certain pesticide product-specific information—such as product chemistry studies related to storage stability that are used to determine label requirements—that do not fall under FIFRA Section 3(c)(7). The incorrect classification of actions as conditional registrations, according to the internal review, may leave the agency vulnerable to allegations by environmental, industry, and other stakeholders who assert that EPA inappropriately grants conditional registrations. However, according to OPP officials, all of the actions that were mistakenly categorized as conditional registrations were legitimate program actions that were lawful under other sections of FIFRA. We were unable to verify this assertion. Further, EPA still needs to take steps to correct these misclassifications in order to ensure the accuracy and integrity of its data and make clear the statutory basis for these program actions.

Several Weaknesses Contributed to Incorrect Data Entry on Conditional Registrations

Several weaknesses contributed to incorrect data entries into OPPIN. First, according to OPP officials, OPP regulatory staff did not have sufficient guidance or training to help them determine when a program action met the criteria for conditional registration. Second, according to OPP's internal review, there was limited, organized management oversight to ensure that regulatory actions not subject to the narrow scope of section 3(c)(7) were not mischaracterized by OPP staff as conditional registrations. As a result, as the internal review stated, the actions that were classified as conditional registrations have varied across OPP's three divisions and by individual entering data into OPPIN within each division. In addition, data management weaknesses contributed to the misclassification of registrations or other actions as conditional. For example, OPP officials said that OPPIN does not generate management reports of summary data that could have alerted managers to the excessive use of "conditional registration" due to the inaccurate classification of actions as conditional registrations by OPP staff. Under the federal standards for internal control, federal agencies are to employ internal control activities, such as management reviews at the functional or activity level, to help ensure that management's directives are carried out and to determine if agencies are effectively and efficiently using resources.[27] Without the ability to generate summary data on conditional registrations from OPPIN, OPP managers cannot easily conduct such reviews. However, they are still responsible for monitoring and ensuring the accuracy of conditional registration data.

In light of the apparent widespread misclassification of regulatory actions as conditional registration, OPP, as part of its internal review of conditional registrations, analyzed data in OPPIN to, among other things, determine EPA's historical use of conditional registrations, including how many of each of the three types of conditional registrations authorized by FIFRA section 3(c)(7) had been granted. The OPP official primarily responsible for conducting this analysis said the intent of the analysis was to show that (1) as noted on the OPP website, only a small portion of the conditional registrations granted by EPA were for new uses under section 3(c)(7)(B) or new active ingredients under section 3(c)(7)(C), as intended, and (2) most of the conditional registrations granted were for identical or substantially similar products under section 3(c)(7)(A). However, in reviewing this information and related documentation, we found that the information on the website was unclear, contained discrepancies, and used technical terms without defining them, which could lead to misinterpretation of the information. For example, the calculations presented on the website to support the conclusion that the overwhelming majority of actions identified in OPPIN as conditional registrations fall outside the circumstances authorized by FIFRA section 3(c)(7) incorrectly grouped conditional registrations for identical or substantially similar products authorized by section 3(c)(7)(A) with label amendments and other actions that fall outside the narrow scope of section 3(c)(7). After meeting with OPP officials in November 2012 to discuss the analysis, these officials acknowledged the website could be clearer and said that the website would be revised to clarify any confusing language and correct any inaccurate statements. However, OPP had no specific plan or time frame for doing so. As of July 2013, these clarifications and corrections had not been made.

OPP Is Taking, or Plans to Take, Actions to More Accurately Account for Conditional Registrations

Accurate and reliable data are essential to an efficient and effective operating environment in the federal government. To more accurately report on the number of pesticide products that are conditionally registered, OPP officials told us that the office has taken or planned to take the following actions:

- Beginning in the fall of 2010, representatives of the OPP divisions that deal with pesticide registrations began meeting with OPP management at least quarterly to, among other things, review proposed conditional registrations for pesticide products with new active ingredients to ensure that (1) any new conditional registrations granted for these products meet the circumstances outlined in FIFRA Section 3(c)(7)(C); (2) the additional data that would be requested as a part of the conditional registration are really needed; and (3) if the data are needed, EPA is still able to make the determination that the information available concerning the pesticide demonstrates that the FIFRA safety standard will be met, which requires that the use of the product during the time needed to generate the necessary data will not cause unreasonable adverse effects on the environment. According to OPP officials, since they started these quarterly reviews, the number of conditional registrations granted for new active ingredients generally has dropped. For example, since starting these reviews, they have been able to preclude cases of misclassification of new active ingredient registrations as "conditional" that had been occurring in the past. Specifically, they noted that, prior to 2010, in some cases, OPP staff had classified some of these registrations as conditional when the additional data being requested of the registrant could only be generated after the date of registration, such as data measuring the storage stability of a commercially manufactured version of the newly registered pesticide product. According to OPP officials, they do not regard such data requirements as being within the scope of FIFRA section 3(c)(7).

- In 2012, OPP began revising the registration categories in OPPIN to, among other things, more accurately reflect those circumstances under which conditional registrations may be granted under FIFRA. As of May 2013, OPP officials said that they had completed development of the categories and provided training to their regulatory staff on how to correctly assign the new categories to each type of registration. In July 2013, OPP officials said they had completed implementation of the new codes in OPPIN. In addition to the training, OPP officials noted the training materials will be available online for regulatory staff to consult for guidance on an ongoing basis.
- In fiscal year 2013, OPP began using a portion of the maintenance fees it collects to begin development of an electronic tracking system for conditional registrations. As discussed, OPP officials are not certain what the total cost of the system will be or when the system will be ready for implementation.

Table 1 summarizes the status, according to OPP officials, of key actions taken or planned by OPP to improve the reliability of conditional registration data.

While OPP officials acknowledged the need to ensure that registrations are accurately classified to reflect their statutory basis and to develop an electronic data system for tracking the status of conditional registrations, they stated that EPA's past practices for managing conditional registrations have not created additional risks to the environment and have been in compliance with applicable laws. Specifically, an EPA attorney stated that the agency views products conditionally registered as identical or substantially similar to currently registered pesticides under section 3(c)(7)(A) and as new uses of currently registered pesticides under section 3(c)(7)(B) as meeting the same safety standards as products registered "unconditionally" under section 3(c)(5). Therefore, according to this official, these products do not pose unreasonable adverse effects on human health or the environment. Further, a Deputy Director of OPP said section 3(c)(7)(A) and section 3(c)(7)(B) registrations make up the bulk of conditionally registered pesticides. In addition, OPP officials stressed that the program actions that were mischaracterized as conditional registrations were nevertheless legitimate program actions. Specifically, these officials and OPP's website note that most of these actions were taken pursuant to the authority of FIFRA implementing regulations and should have been identified as such. Moreover, the EPA attorney stated that FIFRA does not require EPA to convert a registration from "conditional" to "unconditional" when all additional data requirements have been satisfied. So, according to this official, the fact that EPA has not done so in OPPIN in the past does not raise a legal issue. However, despite these assertions, EPA should still take steps to ensure that its data on the registration status of individual pesticide products are current and accurate.

Extent to Which EPA Ensures That Registrants Submit Additional Required Data and EPA Reviews These Data Is Unknown

The extent to which EPA ensures that registrants submit the additional data required when it grants conditional registrations or that it has reviewed these data is unknown. In particular, OPP does not have a reliable system, such as an automated data system, designed specifically to track key information related to conditional registrations, including whether registrants submitted additional data within required time frames and OPP reviewed these data. OPP officials acknowledged this lack of a comprehensive means to track the status of conditional registrations, and noted, as discussed, their intention to develop such a system. However, these officials, as well as OPP's website, note that the conditions for most of these registrations have likely been satisfied as a result of routine program operations that, in the OPP officials' view, constitute a quality assurance check. While these program operations may help to identify some situations where required data are missing, they fall short of what is needed because they are neither comprehensive nor do they ensure the timely submission of these data. This is a key reason that OPP officials are currently conducting a manual review of the files of the more than 16,000 active pesticide registrations OPP has issued, including conditional registrations, to identify any missing data, misclassified registrations, or other problems.

Table 1. OPP Actions Taken or Planned to Improve the Reliability of Conditional Registration Data

Action	Planned completion date	Status	Potential barriers that may affect completion
Conduct quarterly OPP meetings to review the status of conditional registrations.	Ongoing	Began holding meetings in fall 2010.	None
Develop and implement new registration categories in OPPIN to more accurately reflect statutory basis for registration, including conditional registrations.	June 2013	Development of codes for these categories was completed in spring 2013. In July 2013, OPP officials indicated that they had completed the implementation of the new codes in OPPIN.	None
Train regulatory staff to use new codes.	Spring or summer 2013	Completed in spring 2013.	None
Design a new automated system that will include tools to improve the identification, tracking, reporting, and program management of conditional registrations.	2014 or beyond	In fiscal year 2013, began using a portion of the maintenance fees collected to start developing the system.	Complexity of system and sizeneeded to manage the data. Availability of needed funding.

Source: EPA.

OPP's Lack of a Reliable System to Track Conditional Registrations Hinders Its Ability to Ensure that Data Are Submitted, Timely, and Reviewed

OPP lacks a reliable system specifically to track the status of conditional registrations to ensure that additional required data are submitted and timely, and that OPP reviews these data. As discussed, EPA currently "tracks" conditional registrations in OPPIN, an older data system that was not designed for this purpose and that does not have, among other things, the capability to flag situations in which required data have not been submitted by registrants or reviewed by OPP. Federal internal control standards require, in part, that information should be recorded and communicated to management and others within the entity who need it and in a form and within a time frame that enables them to carry out their internal control and other responsibilities.[28] Thus, for an entity to run and control its operations, it must obtain, maintain, and use relevant, reliable, and timely information for program oversight and decision making. Furthermore, the Office of Management and Budget (OMB) directs agency managers to take timely and effective action to correct internal control weaknesses.[29] As measured against this internal control standard, EPA's lack of a reliable and comprehensive means of routinely collecting and tracking information on conditional registrations, including the status of registrants' submission of required data and OPP's review of these data, constitutes an internal control weakness and leaves OPP without an important management tool. For example, when registrants miss due dates without applying for waivers or extensions, it is difficult for OPP without a reliable tracking system to identify these cases for priority follow-up and notify the registrants that their pesticide registrations could be cancelled.

In addition, without a reliable tracking system, OPP may miss conditional pesticides where, had the additional required data been submitted and reviewed, OPP might have altered the terms of a registration. OPP officials acknowledged there have been cases in which their consideration of these additional data led them to make minor changes to a registration, although they could not recall a case where these additional data prompted them to cancel a registration. These officials emphasized, and OPP documents state, that in issuing a conditional registration, even though OPP may ask the registrant for additional data, OPP has determined that the pesticide when used in accordance with labeling and common practices will not cause unreasonable adverse effects on the environment, and that OPP's registration decision takes into account the economic, social, and environmental costs and benefits of the use of that pesticide. Nevertheless, without the ability to systematically track conditional registrations, OPP is not well-positioned to produce summary data to enable it to easily identify situations for priority follow-up; enforce FIFRA and its implementing regulations; and report to Congress and others on program status. For example, without this tracking, it is more difficult to identify patterns of potential problems for management attention, such as registrants that are repeatedly late in providing additional required data for their conditional registrations, which could be the basis for canceling these registrations.

OPP's problems with data management have been well-documented over the years. GAO studies dating back to 1980, 1986, 1991 and 1992 noted problems with OPP data systems used to track the status of pesticide registrations.[30] For example, the 1986 study found that OPP did not have a data system for monitoring whether registrants were submitting the data required by conditional registrations and could only determine the status of data submissions by performing a time-consuming manual file search. At the time, we recommended that OPP take steps to review outstanding conditional registrations of new active ingredients and determine what progress is being made by registrants to submit the required data and take appropriate action. In response to our recommendation, OPP said it was developing a new automated system to track all outstanding data requirements. However, as discussed, OPP does not currently have such a system. The 1992 study noted that, after having spent $14 million over 3 years in data systems development, OPP could not easily assemble accurate, reliable, and complete information on pesticides subject to reregistration (now registration review). EPA Inspector General studies in 1994 and 2000 noted that OPP had not completed actions to improve information systems that contain inaccurate, incomplete, and duplicate data or that are not integrated.[31] In addition, a 2007 EPA contractor study found that many of these problems persist, especially with OPPIN.[32] Noting that OPPIN was launched in 2000,[33] this study found that this system had failed to meet the needs of OPP staff and that many of these staff had created "one-off" (off-line) tracking systems in order to get their jobs done, making comprehensive, reliable status updates, such as whether required data had been submitted and reviewed, very difficult to retrieve. The study also reported that OPPIN lacks the needed data fields and reporting functions for detailed tracking of the status of pesticide registrations, and that multiple OPP staff had expressed dissatisfaction with OPPIN, stating that it is not user-friendly, data are not current or complete, and it lacks a "report card" function to easily check the status of pesticide registrations, including reregistration status. Although the contractor study noted that OPPIN was to be retired in September 2008, we found, as discussed, that OPP is still using OPPIN despite its many limitations.

Although Helpful in Identifying Missing Information, Routine Program Operations Fall Short of What Is Needed

OPP officials said, and the internal review states, that the conditions for most conditional registrations have likely been satisfied as a result of routine program operations that, in their view, constitute a quality assurance check. However, these operations fall short of what is needed for quality assurance because they are neither comprehensive nor do they ensure the timely submission of the additional data required as a condition of the registration. The program operations mentioned by OPP include the following:

- good faith submissions made by registrants to satisfy additional data requirements;
- record keeping and targeted follow-up done by pesticide product managers;
- state pesticide registration actions that may bring to light missing data required by the OPP's conditional registration of a pesticide;
- missing data identified as part of OPP's periodic reevaluation of registered pesticides; and
- registrant-initiated actions, such as label change amendments, that bring to light missing data associated with an earlier conditional registration.

The program operations OPP officials identified may help identify some situations where required data are missing, but they each have limitations and fall short of what is needed, as follows:

- *Registrant submissions:* While undoubtedly many registrants are conscientious about their timely submission of additional data required by their conditional registrations, OPP has found cases in the past where required data were not submitted or were submitted late. As discussed, once OPP issues a conditional registration, the registrant can move the associated pesticide product into the marketplace. In that sense, the registrant's commercialization of that product is not contingent on the registrant's submission of the additional required data. In addition, because OPPIN does not have the ability to systematically flag missing or late data, some registrants may be emboldened to delay or give less priority to developing and submitting these additional data.

- *Product manager actions:* The record keeping and targeted follow-up done by OPP product managers also have limitations. According to OPP officials, each of OPP's 20 product managers is responsible for tracking about 800 of the more than 16,000 active pesticide registrations maintained by OPP. While OPP officials acknowledged that each manager has a very broad span of control, there are other OPP regulatory staff who assist these managers. Moreover, of the approximately 800 registrations handled by each manager, conditional registrations constitute a subset, particularly for new active ingredients. However, we found that OPP had not provided written guidance to product managers on how to track the status of the pesticide registrations for which they are responsible. As a result, according to OPP officials, product managers use a variety of methods to track this information, including electronic spreadsheets or reminder notices, handwritten notes, and memory. Without OPP guidance on how product managers should maintain their pesticide registration files, in the case of the retirement or resignation of an experienced manager—or even the extended absence of a manager due to illness—other managers asked to replace or fill in for this manager may not be familiar with how he or she maintained files or data, or the extent to which this official relied on memory versus written records. Furthermore, requiring all product managers to track the status of registrations in a consistent, electronic format would help OPP meet the goals of an August 2012 OMB directive that, among other things, directs executive agencies to the fullest extent possible to eliminate paper and use electronic record keeping to ensure transparency, efficiency, and accountability. The directive is applicable to all executive agencies and all records.[34]
- *State registration:* While state pesticides registration activities may help to bring to light missing data associated with conditional registrations issued by OPP, the extent to which this happens is unknown and should not be relied upon as a quality assurance check. After OPP registers a pesticide, states also can register that pesticide under specific state pesticide registration laws. A state may be more stringent in registering a pesticide for use in that state, but its

registration requirements generally may not be less stringent than the federal requirements. In addition, states generally have primary responsibility (known as "primacy") for enforcement of the proper use of pesticides within their borders.

- *Periodic review of registered pesticides:* As discussed, FIFRA requires that EPA periodically reevaluate registered pesticides to ensure that each registration continues to satisfy regulatory standards. EPA originally did this reevaluation under its reregistration program, applicable to pesticides registered prior to November 1984. More recently, this reevaluation is being done under the agency's registration review program. According to OPP officials, this periodic review of previously registered pesticides provides an opportunity to identify missing data required by a conditional registration. For example, these officials said any missing data related to section 3(c)(7)(A) (identical or similar) or (B) (new uses) can be identified through registration review. They explained that registrations under these sections do not impose new data requirements. Instead, these registrations are issued when there is an outstanding DCI, or planned DCI, for an identical or similar currently registered pesticide. Thus, according to OPP officials, because registrations under these sections are linked to DCIs, and OPP's Pesticide Registration Information System (PRISM) is used to track DCIs,[35] they are confident in their ability to ensure the timely submission of required data. They noted that PRISM is a newer and more robust data system than OPPIN. They also noted that (1) most conditional registrations are made under sections 3(c)(7)(A) or (B), (2) the associated DCIs place a legal obligation on the registrant to provide the requested data, and (3) OPP may suspend a registration for failure to respond to a DCI. However, we note that PRISM has apparent limitations as well. For example, OPP officials said that PRISM is not designed, per se, to track conditional registrations and therefore cannot be used to identify the conditional registrations, if any, associated with a particular DCI. In addition, they said while this system is useful for tracking the status of DCIs on a case-by-case basis, PRISM lacks the capability to produce summary reports for management attention that could indicate, for example, the extent to which registrants are meeting the requirement to provide an initial response to a DCI within 90 days.[36] Furthermore, as acknowledged by OPP officials, registration review is not helpful in tracking the status of conditional registrations made for new active ingredients under FIFRA section 3(c)(7)(C). For these registrations, there is no relationship to an identical or similar, currently registered pesticide. As a result, the potential exists that a conditionally registered pesticide under this section would continue to be sold and used in the United States for a number of years before OPP discovered that the additional data required had not been submitted and were late. In reviewing OPP documents related to the conditional registration of new active ingredients done in the early 2000s, we noted such cases. In each case, when the associated pesticide product came up for registration review, OPP determined that some of the required data related to the original conditional registration had not been submitted and were late or, if submitted, had not been reviewed by OPP. For example, in the case of a pesticide product containing the active ingredient Foramsulfuron, conditionally registered in November 2002, two required studies on the effects of this pesticide on terrestrial and aquatic plants that were due in December 2004 had not been submitted 10 years after the conditional registration was issued, as determined by OPP's registration review of this pesticide in 2012.[37] In another case, involving a pesticide product containing the active ingredient Acetamiprid, conditionally registered in March 2002, OPP discovered during its registration review of this pesticide in 2012, about 10 years later, that it had received, but not reviewed, a study related to the effects of this pesticide on honeybees. OPP documents indicate the registrant submitted this study in 2001, even before OPP granted the conditional registration. Acetamiprid belongs to a class of pesticides called neonicotinoids that some beekeepers, environmental groups, and others suspect of having adverse effects on honeybees.

- *Registrant-initiated actions:* According to OPP officials, if a registrant applies to amend a registration, such as to make a label change, this action triggers an OPP review of the data supporting the pesticide registration and provides an opportunity to identify missing data associated with a conditional registration. If it is determined that data are missing, the agency may issue a DCI requiring the registrant to provide the data by a specific date. However, this mechanism to identify missing data is ad hoc and only applies to cases in which a registrant seeks to amend its registration.

In addition, OPP officials pointed to an analysis they performed as part of OPP's 2011 internal review that showed registrants usually meet the additional data requirements associated with their conditional registrations on a timely basis. Specifically, OPP examined 544 conditional registrations that it granted (1) for new uses or new active ingredients under FIFRA sections (3(c)(7)(B) or (C), respectively, and (2) from March 2004 to September 2010. According to OPP officials and the office's website, these conditional registrations were selected because, as relatively recent registrations, they were the most likely candidates to be missing additional required data. In contrast, according to OPP officials, older conditional registrations were less likely to be missing data because of the application of the cited routine program operations.

To do this analysis, OPP officials said they first identified all additional data requirements and associated due dates for these conditional registrations, and then manually reviewed the files for these pesticides to determine if the required data were submitted and due dates met. When finished, OPP concluded that registrants had completed 96 percent of "all actions intended" for these 544 conditional registrations in a timely manner. OPP posted this information on its website. However, in reviewing the information on the website, as well as supporting documents provided by OPP, we were unable to verify these calculations, in part because OPP was unable to locate some of the supporting documentation. In addition, some of the documentation provided did not always make clear whether the submitted data had been timely or reviewed by OPP. Furthermore, we found that some of the statements on the website were confusing; used technical terms such as "registration," "action," and "decision" without defining the terms; and contained other discrepancies. Some of these problems have been cited by legal and environmental groups who found this information confusing as well. Although OPP officials generally agreed that revisions were needed to the discussion on the website for clarity, OPP had no specific plan or time frame for doing so. Regardless of any clarifications needed to the website, an OPP Deputy Director expressed confidence in the results of this analysis. As of July 2013, the discussion of this analysis on the website had not been clarified.

OPP Is Manually Reviewing All of the More than 16,000 Pesticide Registrations to Identify Problems Such as Missing Data

In May 2012, OPP staff began manually reviewing the files for the more than 16,000 pesticide registrations granted by EPA. The purposes of this review include identifying all outstanding data requirements, as well as cases where the registration action, such as a label change amendment, was mischaracterized as a "conditional registration." According to OPP officials, all prior registrations are being reviewed, not just those classified as conditional, because pesticide registrations may have long histories, and even though a registration may have been classified initially as unconditional, OPP may have imposed additional data requirements at a later time.

According to OPP officials, reviewing each pesticide registration file is time-consuming and, depending on the pesticide, may take from a few hours to a few days to complete. This generally includes OPP regulatory staff reviewing voluminous paper files associated with many of these pesticide registrations. According to these officials, once this review process is completed, OPP will have a "clean" set of data that will, among other things, identify (1) any missing data, (2) missed deadlines for registrants submitting these data, and (3) cases where the registration action was mischaracterized, including the misuse of "conditional registration." These officials said the review results are being recorded in an electronic spreadsheet, known as the "master file," for future use. For example, when a pesticide comes up for registration review, OPP officials said that staff will refer to the master file to identify any missing data that should be included in the DCI resulting from that review.

Although OPP's review of prior pesticide registrations remains a work in progress, OPP provided an excerpt from its electronic spreadsheet showing the results for three pesticides that we asked about because other agency documents we reviewed suggested possible registration issues. For these pesticides, the spreadsheet indicated that most, but not all, of the required data had been submitted by registrants; some of these data were submitted from 2 to 12 months after the related due dates and, for one of these pesticides, no due date was specified in the registration notice, making it impossible to determine if the data were submitted on-time. In this last case, OPP officials stated that the lack of due dates in the registration notice was an unintentional oversight. Since OPP has not finished its review of all prior registrations, it has not yet developed summary statistics on the frequency with which these types of problems were found. OPP officials said they plan to complete their manual review of prior registrations by the fall of 2013, but its completion by then will depend on the amount of time OPP staff can devote to this review relative to their other responsibilities. In addition, according to agency officials, by the end of calendar year 2013, they plan to make public information from the review about the number of active ingredients approved under FIFRA section 3(c)(7)(C) for which data are overdue, and those for which data were submitted late.

While OPP's manual review of existing registrations may result in a clean data set and identify some missing data and other problems not discovered as a result of what OPP calls routine program operations, it is an interim measure. Among other things, OPP officials said their office needs a comprehensive automated data system for tracking conditional registrations. As noted by one OPP division director, "no one wants to have to track this information by hand" in the future.. In addition, according to these officials, OPP does not plan to update the master file to include new pesticide registrations or other registration changes that occur in the future. They noted that the master file is retrospective, and it provides a snapshot in time. Instead, OPP officials said that new pesticide registrations and other registration changes will be entered into OPPIN using the new codes that OPPIN plans to introduce in June 2013. In July 2013, EPA advised us that the codes have been entered in OPPIN. According to these officials, the introduction of these new codes, in conjunction with staff training on how to use them, should preclude some of the misclassification problems experienced in the past until a new comprehensive data system is available to replace OPPIN.

Stakeholders Generally Said EPA Needs to Improve Its Conditional Registration Process and Had Mixed Views on the Value of These Registrations

The 24 stakeholders that responded to our questionnaire—including representatives of consumer (3), environmental (6), industry (5), legal (3), producer (1), science (1), and state government groups (5)—generally indicated that EPA needs to improve its conditional registration process and, in some cases, they offered suggestions for improving this process.[38] The issues that stakeholders raised included concerns about the timely submission and review of required data and the misuse of the conditional registration designation. In addition, stakeholders' views varied regarding the potential benefits and problems associated with the conditional registration of pesticides.

Respondents Generally Reported That EPA Needs to Improve Its Conditional Registration Process

In responding to our questionnaire, respondents in the consumer, environmental, industry, legal, science, and state government stakeholder groups generally reported concerns with submission or review of required data as follows:

- Of the 19 respondents in the consumer, environmental, legal, producer, science, and state government groups, 17 reported concerns related to registrants not submitting additional required data on time, including concerns about pesticides that remain in the marketplace when their environmental and health impacts have not been fully evaluated.[39] In addition, 3 of the 8 respondents from consumer groups and state government were generally concerned that, when registrants are allowed to miss due dates without any follow-up from EPA, there is little incentive for the registrants to submit the additional data and take the data requirements seriously. Further, 4 of the 17 respondents from consumer, environmental, legal, and state government groups reported that EPA should cancel registrations for those registrants who do not submit required data on time.
- Of the 19 respondents from environmental, industry, legal, and state government groups, 8 reported concerns about EPA's record keeping related to conditional registrations, including the agency's ability to ensure the receipt and review of required data. Specifically, one industry stakeholder stated that EPA does not effectively track the receipt and review of required data and said, in particular, EPA does not always acknowledge receipt of required information and does not always notify recipients whether the data submitted satisfied the condition.[40] However, 4 of the 5 respondents from industry stated the public faced little risk when required data are not submitted or reviewed, pointing out that before EPA conditionally registers a pesticide, the agency must determine that the pesticide meets FIFRA registration standards.

Stakeholders representing environmental, industry, and state government groups generally reported concern with EPA issuing conditional registrations for circumstances that were outside of the permissible situations stated in FIFRA. Examples are as follows:

- Of the 16 respondents representing these groups, 10 reported concern that EPA was overusing conditional registrations. Three of the 5 environmental stakeholders that had these concerns generally stated that conditional registrations were originally intended to be used in limited circumstances where a public need was established— such as the need to quickly approve the use of a pesticide to prevent significant crop damage and economic loss—and that EPA's current practices for issuing conditional registrations are not in keeping with this original intent.[41]
- Of the 5 respondents from industry, 3 stated that there are cases when EPA grants a conditional registration for a pesticide that should have qualified for an unconditional registration. These stakeholders reported that, in some of these cases, EPA had granted a conditional registration for reasons that were outside of those permissible circumstances outlined in FIFRA, such as the need for labeling changes and potential data requirements that could be imposed in the future as a result of the registration review process. Of these three industry stakeholders, two were aware of EPA's recent efforts to ensure that conditional registrations are granted only in appropriate circumstances, as described on the agency's website, and these stakeholders were supportive of these efforts.

In keeping with these concerns, respondents from environmental, industry, producer, and state government groups generally offered suggestions for improving EPA's conditional registration process. Following are examples:

- Of the 16 respondents representing environmental, industry, and state government groups, 7 generally stated that EPA needs a better system for tracking the status of conditional registrations, including its review of required data.
- Of the 10 respondents representing industry and state government groups, 6 stated that it would be helpful if EPA developed a way to share information about conditional registrations with external stakeholders. For example, a respondent from a state government

group suggested that EPA create a notification and tracking system for states that specifically lists the status of conditional registrations and any pending data requirements. The stakeholder stated that this system would facilitate the exchange of information between EPA and the states. Also, a respondent representing an industry group stated that EPA should develop a system similar to one used by California's Department of Pesticide Regulation; the respondent stated that this system allows registrants to access information on the status of their pesticide registration applications pending before that state and about their conditionally registered products.

- Of the 5 industry respondents, 2 suggested that EPA address concerns about overuse of conditional registrations by taking steps to ensure that these registrations only are used according to the explicit criteria set forth in FIFRA section 3(c)(7).

Some Respondents Stated That Conditional Registrations Offer Benefits, and Others Stated That They Pose Serious Risks

Of the 17 respondents representing consumer, industry, legal, producer, and state government groups, 14 stated that there were significant benefits in conditionally registering pesticides. They generally noted that the conditional registration process is an important and effective mechanism that gives EPA the flexibility to allow pesticides to move to the market more quickly, and that quicker movement to the market, in turn, provides users and growers with faster access to the pesticides they need. For example, one respondent commented that conditional registrations are especially valuable in situations where users have a much more limited choice of pesticides that meet their needs, such as growers of specialty crops.[42] Seven of these 14 respondents also generally stated that conditional registrations promote innovation by bringing new technologies and products to the marketplace faster.

In contrast, 13 of the 18 respondents representing consumer, environmental, legal, science, and state government groups stated that there were numerous negative impacts caused by conditionally registering pesticides. For example, all 13 of the respondents concerned with negative impacts reported that conditionally registering pesticides can delay EPA's ability to mitigate public health and environmental impacts caused by pesticides. These respondents generally stated that EPA is not conducting a full, rigorous review of conditionally registered pesticides and therefore is allowing these pesticides into the marketplace without complete data, such as toxicity tests and studies that demonstrate the pesticides' impact on the environment. Eight of these 13 respondents who expressed concern with conditionally registering a new use of a pesticide generally noted that, without a full, rigorous review, EPA may miss problems caused by the new use that may not have occurred with the original use.[43] Four of the 13 respondents also stated that, especially in the cases that involve pesticides with new active ingredients, a conditional registration should only be granted if a critical need for the pesticide can be demonstrated. Six of the 13 respondents stated that the risks posed by conditionally registering new active ingredients were so great that EPA should discontinue this type of conditional registration. In elaborating on their concerns, 12 of the 18 respondents from consumer, environmental, legal, science, and state government groups cited examples of conditionally registered pesticides that, in their opinion, should not have been conditionally registered. The three pesticide products mentioned most frequently by these respondents were the following:

Clothianidin

Background: According to EPA documents, in 2003, EPA conditionally registered the insecticide Clothianidin, which the agency had identified as an alternative to older, more toxic insecticides. As one of the conditions of the registration, EPA required the manufacturer, Bayer CropScience, to submit a study evaluating the effects on honeybees of prolonged exposure to Clothianidin. In 2007, the agency reviewed this study and determined that it satisfied EPA's field study guidelines. However, in 2010 and again in 2012, numerous entities, including consumer and environmental groups, petitioned EPA to discontinue use of Clothianidin, charging, among other things, that it posed an imminent hazard to honeybees. In 2010, EPA decided to reevaluate the study and ultimately determined that there were some deficiencies in the study but that the registered uses of Clothianidin met the FIFRA standard for registration. According to Bayer CropScience, the use of Clothianidin was necessary to prevent crop damage from pesticide-resistant pests and the dying of honeybees in 2008, which environmental groups claimed was caused by Clothianidin, was actually a result of a variety of factors, including incorrect application of the pesticide.

Stakeholder comments: Half (9 of 18) of the respondents representing consumer, environmental, legal, science, and state government groups raised concerns about the insecticide Clothianidin. For example, of the 9 respondents who raised concerns, 6 generally reported that EPA should not have conditionally registered this pesticide in 2003 without all the data needed to establish that the pesticide would not significantly increase unreasonable adverse effects to pollinators, including honeybees. Two of these 6 respondents reported that the failure to do so has allowed the widespread use of a pesticide that, in their view, has caused the death of honeybee colonies and irreparably damaged the environment and livelihoods of beekeepers. Four of the 6 respondents said that it is irresponsible for EPA to refuse to discontinue the registration of this product when the agency eventually determined that the pollinator study submitted by the company was inadequate.[44]

Imprelis

Background: According to EPA documents, in August 2010, EPA conditionally registered the active ingredient of the pesticide Imprelis. This pesticide, manufactured by DuPont, was a low-toxicity herbicide used to control weeds, vines, and grasses on nonfood use sites, such as weeds around an office building. EPA stated that the studies originally submitted for Imprelis were adequate to make a finding for the registration but also concluded that two additional studies (on toxicity and reproduction) were required to confirm the conclusions from existing data. However, in the summer of 2011, EPA received reports from several states that this pesticide may have caused injury to certain species of evergreen trees, particularly Norway spruce and white pine. In a June 2011 letter, DuPont cautioned professional applicators not to use Imperils near certain species of trees, including Norway spruce and white pine. On August 4, 2011, DuPont voluntarily suspended sales of Imprelis and, on August 11, 2011, EPA issued a stop-sale order directing DuPont to immediately halt the sale, use, or distribution of Imprelis. According to EPA, it issued this stop-sale order because it had reason to believe the product was misbranded and the agency had obtained new information, not available during the registration process, that showed Imprelis was toxic to certain trees. Currently, EPA is evaluating the tree damage to determine what caused the injuries. DuPont has since started a return and refund program for Imprelis users.

Stakeholder comments: Of the 7 respondents representing environmental and science groups, 4 raised concerns about the pesticide Imprelis. According to 3 of these 4 stakeholders, EPA did not properly consider the "unreasonable effects on the environment" of this pesticide, specifically effects on organisms that this pesticide was not intended kill, including trees. One stakeholder reported that the experience with this pesticide illustrates what can happen when EPA allows a pesticide product to proceed to the marketplace without complete information and, in this stakeholder's opinion, confirms that the conditional registration process actually allows EPA to bypass statutory safeguards and rush pesticides with unknown and unevaluated risks to market.[45]

Nanosilver

Background: According to EPA documents, in December 2011, EPA issued a conditional registration for a pesticide product containing nanosilver as a new active ingredient. As we reported in May 2010, nanosilver is one of a number of nanomaterials, which are materials that occur at the scale of nanometers—the equivalent of one-billionth of a meter, increasingly being used in commercial and industrial products.[46] The antimicrobial pesticide product, HeiQ AGS-20, is a silver-based nanomaterial for use as a preservative for textiles. As a condition of registration, EPA is requiring the registrant, HeiQ Materials AG, to submit a number of studies to confirm the agency's assessment that the product will not cause unreasonable adverse effects on the environment. These studies include toxicity studies for occupational, residential, and environmental exposure scenarios.[47] The data must be submitted within 4 years (i.e., by December 2015) to avoid cancellation of the conditional registration. The registrant agreed with the agency's decision to grant a conditional registration. The registrant stated that this registration was appropriate because (1) the company had insufficient time to generate certain required data, (2) use of the pesticide is in the public interest, and (3) use of the pesticide during the period needed to generate and review the required data will not cause unreasonable adverse effects. The registration of this product is being challenged in a lawsuit by the Natural Resources Defense Council.[48]

Stakeholder comments: Eight of the 18 respondents representing consumer, environmental, legal, science, and state government groups raised concerns about the conditional registration of the nanosilver product HeiQ AGS-20. Among other things, these respondents stated that there was a significant lack of data needed to fully evaluate the safety of this product, including a lack of data on potential human health effects. One stakeholder pointed out that silver is highly toxic to both humans and the environment and, when used at the nanoscale level, the science is still developing on how such nanoparticles behave. Another stakeholder said this conditional registration is particularly troubling because EPA's Office of the Inspector General recently concluded that the EPA lacks an effective program to collect information and monitor the possible health risks posed by nanomaterials,[49] and because the National Research Council similarly found that the federal government needs an improved plan and additional funding to assess the environmental and health risks posed by nanomaterials.[50]

CONCLUSION

OPP faces a formidable task in managing the registrations of more than 16,000 pesticide products. This includes periodically reconsidering whether older registrations continue to meet the regulatory standard and reviewing applications submitted each year for new or amended registrations. The pesticide products that the agency registers play a critical role in food production by helping to minimize crop losses due to pests and weeds. In addition, pesticide products have helped improve public health by controlling disease-carrying pests, such as insects and rodents. At the same time, consumers rely on OPP to ensure that registered pesticide products do not cause unreasonable adverse effects on the environment or human health when used according to the label instructions approved by the agency.

OPP faces challenges in tracking key information specifically related to conditional registrations, and, as a result, is unable to produce accurate information on the current number of these registrations. OPP's lack of a reliable and comprehensive means of routinely collecting and tracking information on conditional registrations, including the status of registrants' submission of required data and OPP's review of these data, leaves it without an important management tool. Because OPP is not systematically tracking whether registrants of conditionally registered pesticides submitted additional required data, and whether OPP reviewed these data, it may not be able to identify situations in which the additional data would suggest the need to alter a registration. Furthermore, without the ability to systematically track conditional registrations, OPP is not well-positioned to produce summary data to enable it to easily identify situations for priority follow-up; enforce FIFRA and its implementing regulations; and report to Congress and others on program status.

Furthermore, OPP's use of conditional registrations for actions other than those that meet the criteria outlined in FIFRA Section 3(c)(7) has created confusion for its staff, and may leave OPP vulnerable to charges by environmental, industry, and other stakeholders who assert that it inappropriately grants conditional registrations. According to OPP officials and documents, weaknesses in guidance and training, management oversight, and data management contributed to the misclassification of other pesticide-related activities as conditional registrations. OPP has raised the visibility of these issues by holding meetings at least quarterly with representatives of the OPP divisions that issue pesticide registrations to discuss any conditional registrations being considered for new active ingredients to ensure they meet the statutory criteria outlined in FIFRA Section 3(c)(7) and that the additional information requested is indeed needed to unconditionally register the pesticide. In addition, OPP has taken or plans to take other actions to ensure that staff appropriately grant conditional registrations and registrants submit required data; however, these are short-term solutions that do not fully address the problems identified. As OPP officials and stakeholders recognize, OPP needs a comprehensive automated data system to track conditional registrations from the time the conditional registration is granted until additional data requested is received and reviewed. However, OPP previously stated over 25 years ago that it planned to develop an automated system for tracking conditional registrations of new active ingredients, but it did not follow through with this plan. OPP has secured funding through FIFRA amendments to begin the development of such a system, but much work remains to be done and will depend on a further commitment of needed resources.

Moreover, OPP lacks written guidance or a consistent methodology for how product managers are to maintain their pesticide registration files. Allowing product managers to use disparate methods to collect and keep information on the pesticides they are responsible for makes it more difficult to develop summary information about the status of pesticide registrations overall, which could be useful information for managing the pesticide registration program. Also, as the extensive amount of time and effort needed in OPP's ongoing review of all registered pesticides demonstrates, product managers' current methods are not sufficient to efficiently track this information. It will take time and money to develop an automated system for tracking the progress of conditionally registered pesticides but, in the interim, OPP is further hampered in its efforts to be informed about the status of conditional registrations because product managers do not use a consistent system for tracking their status, including when data are submitted by registrants and reviewed by OPP. Furthermore, OPP's reliance on the institutional knowledge of its product managers and the records they keep is problematic for other reasons, including the loss of knowledge and experience of these employees as they retire or are replaced by new employees.

Finally, OPP has also not clearly and concisely communicated the results of its analyses of conditional registrations on its website. It is important for agencies to ensure that information placed on their websites is accurate and free of discrepancies. OPP officials have acknowledged the need to correct the agency's website on conditional registrations and remove any confusing or inaccurate statements.

Recommendations for Executive Action

To improve EPA's management of the conditional registration process, we recommend that the Administrator of EPA direct the Director of the Office of Pesticide Programs to take the following three actions:

- Complete plans to automate data related to conditional registrations to more readily track the status of these registrations and related registrant and agency actions and identify potential problems requiring management attention,
- Pending development of an automated data system for tracking the status of conditional registrations, develop guidance to ensure that product managers use a uniform methodology to track and document this information, including when data are submitted by registrants and reviewed by EPA, in the files maintained by each pesticide product manager.
- Review and correct, as appropriate, OPP's website on conditional registrations to ensure that the information presented is clear, concise, and accurate, including defining technical terms.

Agency Comments and Our Evaluation

We provided a draft of this report to EPA for review and comment. In written comments, EPA agreed with the report's recommendations. Regarding the first recommendation, EPA said that its implementation plan to automate data related to conditional registrations includes (1) development of new codes for identifying conditional registration decisions in OPPIN; (2) training its staff on use of the new categories represented by these new codes, and making the training available online for guidance; and (3) changes to its databases to allow staff to check more easily whether there are any outstanding requests for data on any pesticide active ingredients. EPA also said it plans to develop a more comprehensive system for tracking conditional registrations; however, the agency's ability to do so depends on the availability of funding and the complexity of incorporating changes in the databases. Regarding the second recommendation, EPA said it is developing a standard operating procedure for staff to follow when entering data into the computerized tracking system about the statutory basis for registration decisions. According to EPA, this procedure, together with the new training for staff, should ensure that conditional registration decisions are properly identified in the OPPIN database going forward. The agency said it expects to complete the procedure by the end of calendar year 2013. EPA added that the status of products previously approved under conditional registration authority is also being reviewed and updated, as necessary. For the third recommendation, EPA said by the end of 2013 it will revise its website on conditional registration to both clarify and update the information presented. In addition, EPA indicated the website will include an outline of ongoing agency work to strengthen its conditional pesticide registration program. EPA also provided technical comments, which we incorporated as appropriate.

Sincerely yours,

J. Alfredo Gómez
Director, Natural Resources and Environment

APPENDIX I: OBJECTIVES, SCOPE AND METHODOLOGY

Our objectives were to examine the (1) number of conditional pesticide registrations the Environmental Protection Agency (EPA) has granted and the basis for granting these registrations; (2) extent to which EPA ensures that registrants submit the additional data EPA required as part of conditional registrations and reviews these data; and (3) views of relevant stakeholders on EPA's use of conditional registrations, including ways, if any, to improve the conditional registration process. To address these objectives, we reviewed relevant federal statutes and regulations,[51] EPA program and guidance documents,[52] federal internal control standards,[53] and previous GAO and EPA Inspector General reports.[54] We also reviewed EPA's fiscal year 2011–2015 strategic plan; EPA's fiscal year 2011 and 2012 annual performance plans; EPA's budget justification documents for fiscal years 2012 and 2013; Office of Pesticide Programs' (OPP) pesticide registration work plans for fiscal years

2001 through 2012; OPP notices of pesticide registration for conditionally registered pesticides; and Federal Register notices related to OPP's registration decisions. In addition, we interviewed OPP officials and reviewed documentation they provided to obtain further information and clarification on EPA's conditional registration process, including any planned responses to internal review or external stakeholder concerns. Furthermore, we reviewed recent literature related to pesticide registration, including information and documents found on the websites of a variety of consumer, environmental, industry, legal, producer, science, and state government organizations.

To examine the number of conditional pesticide registrations EPA has granted and the basis for them, we requested that OPP provide us with summary data on (1) the number of pesticide registrations currently in conditional status and how long they have been in this status; (2) the total number of current pesticide registrations (conditional and unconditional); and (3) for fiscal years 1997 through 2011, the number of conditional registrations granted for each year and the basis on which these were granted. We asked for information for 1997 through 2011 because reviewing registrations from this period could address the concerns that environmental and other groups raised that some pesticides may have been in conditional status for many years and also take into account key changes made to the pesticide registration and tolerance setting processes that occurred after 1996.[55] We intended to assess the reliability of the data that we requested from EPA and conduct electronic testing on data fields necessary for our analysis; however, after interviewing OPP officials and reviewing past GAO, Inspector General, and EPA contractor studies examining OPP's data management, especially its use of the Office of Pesticide Programs Information Network (OPPIN), we concluded that EPA could not provide us with sufficiently reliable data for obtaining summary level information on conditional pesticide registrations. In the absence of these data, we discussed with OPP officials the capabilities and limitations of OPPIN and any potential work-arounds they employ or are planning for. In addition, we reviewed and discussed with these officials a recent analysis of OPPIN data that OPP conducted to determine how many of the conditional registrations granted were outside of the permissible circumstances outlined in the Federal Insecticide, Fungicide, and Rodenticide Act (FIFRA). The results of this analysis were included in an OPP internal review report and posted on OPP's website.

To address the second objective—the extent to which EPA ensures that registrants submit the additional data required as part of conditional registrations and reviews these data—we interviewed OPP officials about how they manage conditional registration data requirements. We also asked OPP to provide information on the number of pesticides that (1) have a conditional registration, but the registrant has not submitted the additional required data by the specified due date; (2) have a conditional registration and the registrant has submitted the additional required data, but EPA has not reviewed these data; (3) still have a conditional registration, even though the registrant has submitted, and EPA has reviewed, the additional required data; and (4) had been changed from conditional to unconditional status. However, OPP officials said they could not provide these data because they do not have an automated data system that tracks this information, but that they did analyze a subset of 544 conditional registrations to try to determine whether registrants had submitted the required data and posted the results of this analysis on OPP's website.

To obtain the views of relevant stakeholders on EPA's conditional registration process and ways, if any to improve it, we administered a questionnaire to 35 professionals in the consumer, environmental, industry, legal, producer, science, and state government fields. We used a multistage process to identify our final nonprobability sample of 35 potential respondents.[56] This process included (1) conducting a literature search to identify groups or individuals who had recently published articles on registration (including conditional registration) of pesticides, (2) asking agency and other relevant officials for recommendations of knowledgeable parties in each of these areas, and (3) asking prospective stakeholders for suggestions of other potential stakeholders. Through these methods, we arrived at an initial list of 148 potential stakeholders that were divided, based on their institutional affiliation, into the seven categories listed above. We then narrowed this list by (1) performing online searches of the professional affiliations of each stakeholder to determine whether they would likely have sufficient knowledge about EPA's conditional registration of pesticides and (2) conducting screening interviews, by phone and e-mail, to determine whether the individuals were sufficiently familiar with EPA's process for registering pesticides and to secure their commitment to participate in our survey. After this process was completed, we arrived at a list of 35 stakeholders to whom we sent, via e-mail, our questionnaire.

The questionnaire asked about, among other things, (1) problems, if any, associated with each of the permissible situations for which registrations can be conditionally granted; (2) risks, if any, associated with registrants not submitting data required by a conditional registration in a timely manner; (3) pesticides with conditional registrations that stakeholders believe should not have been conditionally registered; and (4) suggestions for improving EPA's conditional registration process. In preparing to administer this questionnaire, we conducted three pretests to ensure that the questions were clear, terminology was used appropriately, and the questionnaire was unbiased. We used the results of our pretests to revise the questions as needed. The questions were open-ended, and thus issues raised by stakeholders had to be "volunteered." We did not ask each stakeholder to agree or disagree with particular issues.

The administration period for this questionnaire was from July through September 2012. Of the 35 participating stakeholders, 24 provided complete, valid questionnaire responses.[57] The organizations that participated as stakeholders were Akin Gump, LLP; American Chemistry Council; American Farm Bureau Federation; Beyond Pesticides; California Department of Pesticide Regulation; Center for Biological Diversity; Center for Environmental Health; Center for Food Safety; Center for Science in the Public Interest; Council of Producers and Distributors of Agrotechnology; CropLife America; Dow Agrosciences, LLC; Earthjustice; Environmental Working Group; Food and Water Watch; Florida Department of Agriculture and Consumer Services; Iowa Department of Agriculture; McDermott, Will and Emery, LLP; Natural Resources Defense Council; New York State Department of Environmental Conservation; Pesticide Action Network North America; Syngenta Crop Protection, LLC; Texas Department of Agriculture; and The Endocrine Disruption Exchange. The results of this questionnaire cannot be generalized to all parties knowledgeable about the conditional registration of pesticides; rather our analysis of the results of this questionnaire identifies common themes present in the responses of those who participated in our questionnaire.

We analyzed stakeholder responses to the questionnaire to identify themes and develop summary findings. Two GAO analysts separately conducted this analysis and placed users' responses into one or more categories, then compared these analyses. All initial disagreements regarding the categorizations of stakeholders' responses were discussed and reconciled. The analysts then tallied the number of responses in each category.

We conducted this performance audit from September 2011 to August 2013 in accordance with generally accepted government auditing standards. Those standards require that we plan and perform the audit to obtain sufficient, appropriate evidence to provide a reasonable basis for our findings and conclusions based on our audit objectives. We believe that the evidence obtained provides a reasonable basis for our findings and conclusions based on our audit objectives.

RELATED GAO PRODUCTS

Nanotechnology: Improved Performance Information Needed for Environmental, Health, and Safety Research. GAO-12-427. Washington, D.C.: May 21, 2012.

Agricultural Chemicals: USDA Could Enhance Pesticide and Fertilizer Usage Data, Improve Outreach, and Better Leverage Resources. GAO-11-37. Washington, D.C.: November 4, 2010.

Pesticides on Tobacco: Federal Activities to Assess Risks and Monitor Residues. GAO-03-485. Washington, D.C.: March 26, 2003.

Agricultural Pesticides: Management Improvements Needed to Further Promote Integrated Pest Management. GAO-01-815. Washington, D.C.: August 17, 2001.

Children and Pesticides: New Approach to Considering Risk is Partly in Place. GAO/HEHS-00-175. Washington, D.C.: September 11, 2000.

Pesticides: Improvements Needed to Ensure the Safety of Farmworkers and Their Children. GAO/RCED-00-40. Washington, D.C.: March 14, 2000.

GAO Products Related to Pesticide Regulation. GAO/RCED-95-272R. Washington, D.C.: September 26, 1995.

Pesticides: Information Systems Improvements Essential for EPA's Reregistration Efforts. GAO/IMTEC-93-5. Washington, D.C.: November 23, 1992.

Pesticides: EPA's Information Systems Provide Inadequate Support for Reregistration. GAO/T-IMTEC-92-3. Washington, D.C.: October 30, 1991.

Pesticides: EPA's Formidable Task to Assess and Regulate Their Risks. GAO/RCED-86-125. Washington, D.C.: April 18, 1986.

Natural Resources and Environment: Delays and Unresolved Issues Plague New Pesticide Protection Programs. GAO/CED-80-32. Washington, D.C.: February 15, 1980.

End Notes

[1] Act of June 25, 1947, ch. 125, 61 Stat. 163 (codified as amended at 7 U.S.C. §§ 136-136y).

[2] The phrase "unreasonable adverse effects on the environment" is defined in FIFRA, in part, to mean (1) any unreasonable risk to man or the environment, taking into account the economic, social, and environmental costs and benefits of the use of any pesticide, or (2) a human dietary risk from residues that result from a use of a pesticide in or on any food inconsistent with the standard for tolerance under section 408 of the Federal Food, Drug, and Cosmetic Act, 21 U.S.C.§ 346a. This standard requires EPA to consider both the benefits and risks of using a pesticide.

[3] According to Office of Pesticide Programs (OPP) officials, applicants for pesticide registrations are usually pesticide product manufacturers. When EPA registers an applicant's pesticide product, the applicant is then called a registrant. For the purposes of this report, we refer to applicants as registrants.

[4] Under the Federal Food, Drug, and Cosmetic Act, EPA is authorized to establish tolerances—or exemptions for the requirement of a tolerance—for pesticide residues that remain in food or animal feed.

[5] An active ingredient is the chemical or substance component of a pesticide product that will prevent, destroy, repel, or mitigate a pest.

[6] EPA refers to registrations that are in effect (i.e., the registration has not been suspended or canceled) as "active" registrations. For purposes of this report, unless otherwise indicated, all registrations discussed are "active."

[7] EPA Office of Inspector General, *EPA's Pesticide Program*, Report no. E1EPE2-05-0015-4100205, Mar. 11, 1994; EPA Office of Inspector General, *Pesticides: Follow-up Report on EPA's Pesticide Program*, Report no. 00P00011, March 27, 2000. GAO reports are listed under "Related GAO Products" at the end of this report.

[8] Because this was a nonprobability sample, the information collected in response to the questionnaire about EPA's use of conditional registrations cannot be generalized to all consumer, environmental, industry, legal, producer, science, and state government stakeholders but provides illustrative information on the views of such stakeholders by category.

[9] 21 U.S.C. §§ 301-399f.

[10] 40 C.F.R. pts. 152-180.

[11] Nonoccupational exposures are those experienced by the general population, as opposed to those experienced by specific groups of pesticide users, such as farm workers and pest control operators.

[12] To make the safety finding, EPA considers, among other things, the toxicity of the pesticide and its breakdown products, aggregate exposure to the pesticide in foods and from other nonoccupational sources of exposure, and any special risks posed to infants and children. EPA may grant an exemption from the requirement to have a tolerance when there is a reasonable certainty that no harm will result from aggregate exposure, including all anticipated dietary exposures for which there is reliable information.

[13] The Registration Division is responsible for certain regulatory activities related to market entry of new pesticides—including product registrations, amendments, tolerances, experimental use permits, and emergency exemptions—associated with conventional pesticides, which are man-made chemicals developed and produced primarily or only for use as pesticides. The Biopesticides and Pollution Prevention Division is responsible for all regulatory activities associated with biologically based pesticides, which are chemicals derived from plants, fungi, bacteria or other natural substances that can be used for pest control, as well as for Plant-Incorporated Protectants, which refers to the material added to or generated in plants that have been genetically engineered to express a pesticidal property. The Antimicrobials Division is responsible for all regulatory activities associated with

antimicrobial pesticides, which are chemical substances that can be used to kill microorganisms.

[14] FIFRA section 3(c)(5) provides that the EPA Administrator shall register a pesticide if the Administrator determines, among other things, that (1) the pesticide's composition is such as to warrant the proposed claims for it; (2) its labeling and other material required to be submitted comply with the requirements of FIFRA; (3) it will perform its intended function without unreasonable adverse effects on the environment; and (4) when used in accordance with widespread and commonly recognized practice it will not generally cause unreasonable adverse effects on the environment.

[15] In addition, FIFRA implementing regulations state that EPA will not approve the conditional registration of a pesticide product for a new use if (1) the pesticide is the subject of a special review to determine if the use of the pesticide may result in unreasonable adverse effects based on a use of the product that results in human dietary exposure or (2) the proposed new use is for a major food or feed crop (such as corn or soy beans), or involves use on a minor food (including many fruits and vegetables) or feed crop for which there is an effective alternative registered pesticide that does not meet the FIFRA criteria for risk associated with human dietary exposure.

[16] In addition to an EPA registration, registrants must also comply with state regulations, some of which require registration in the state before the product can be sold or distributed in that state.

[17] 74 Fed. Reg. 11,551 (Mar. 18, 2009).

[18] A nematicide kills nematodes, which are microscopic, worm-like organisms that feed on plant roots.

[19] FIFRA section 3(g) requires that EPA periodically reevaluate registered pesticides.

[20] Procedural regulations for the registration review of pesticides became effective on October 10, 2006. By law, the agency must complete the first 15-year cycle of registration review by October 1, 2022. The agency must complete the registration review of each new pesticide active ingredient within 15 years of its initial registration. Previously, EPA conducted similar reviews of pesticides first registered before November 1984 under its reregistration program. Registration review is replacing the reregistration program. Unlike reregistration, registration review is to operate continuously and encompass all registered pesticides, according to EPA.

[21] The Pesticide Registration Improvement Act of 2003 (PRIA), Pub. L. No. 108-199, Div. G, Tit. V, § 501, 118 Stat. 419 (2004), amended FIFRA by, among other things, establishing pesticide registration fees for some registration actions. PRIA was reauthorized in 2007 (Pesticide Registration Improvement Renewal Act of 2007, Pub. L. No. 110-94, 121 Stat. 1000 (2007) or PRIA 2) and again in 2012 (Pesticide Registration Improvement Extension Act of 2012, Pub. L. No. 112-177, 126 Stat. 1327, or PRIA 3).

[22] EPA, *Fiscal Year 2012 Justification of Appropriation Estimates for the Committee on Appropriations,* EPA-190-R-11-003.

[23] The Natural Resources Defense Council (NRDC) requested the information regarding EPA's use of conditional registrations.

[24] The groups that concurred with NRDC's position on EPA's use of conditional registrations were Alaska Community Action on Toxics, Beyond Pesticides, Center for Environmental Health, Friends of the Earth U.S., Organic Consumers Association, Pesticide Action Network North America, and TEDX The Endocrine Disruptor Exchange.

[25] OPP's website can be found at http://www.epa.gov/pesticides/regulating/conditional-registration.html.

[26] The Pesticide Registration Improvement Extension Act of 2012 amended FIFRA to provide, among other things, funding to improve the information systems capabilities for OPP; the amendments provide that this funding is to support enhancing the systems capacity to track pesticide registration decisions, including the status of conditional registrations. The amendments authorize $800,000 per year for 5 years to be set aside from the maintenance

fees to fund these enhancements, including those related to conditional registrations. In addition, the amendments require, in part, that EPA report each year through 2017 on its progress in implementing a system for tracking the status of conditional registrations.

[27] GAO, *Standards for Internal Control in the Federal Government,* GAO/AIMD-00-21.3.1 (Washington, D.C.: November 1999).

[28] GAO/AIMD-00-21.3.1.

[29] Office of Management and Budget, *Management's Responsibility for Internal Control*, OMB Circular No. A-123, Dec. 21, 2004.

[30] These reports are listed under "Related GAO Products" at the end of this report.

[31] EPA Office of Inspector General, *EPA's Pesticide Program,* Report no. E1EPE2-05-0015-4100205, Mar. 11, 1994; EPA Office of Inspector General, *Pesticides: Follow-up Report on EPA's Pesticide Program*, Report no. 00P00011, Mar. 27, 2000.

[32] EPA, *Evaluation of the U.S. EPA Pesticide Product Reregistration Process: Opportunities for Efficiency and Innovation*, March 2007, done by Abt Associates under EPA contract EP-W-04-023.

[33] According to EPA, although work on OPPIN began in 2000, the system was not operational until 2003.

[34] OMB, Managing Government Records Directive, M-12-18, Aug. 24, 2012.

[35] PRISM is an OPP data system that is intended to provide a centralized source of information on all registered pesticide products, including chemical composition, toxicity, name and address of registrant, brand names, registration actions, and related data.

[36] In July 2013, in its technical comments on a draft of this report, EPA stated that PRISM is now capable of producing summary reports for management. However, the agency neither explained what these reports are nor did it provide examples.

[37] In July 2013, in its technical comments on a draft of this report, EPA said that after contacting the registrant and determining that the registrant did not intend to fulfill the requirements of its conditional registration for this product, the agency cancelled the registration in February 2013 (78 FR 8513, Feb. 6, 2013). The registrations of other pesticide products containing Foramsulfuron were not affected by this cancellation.

[38] Not every stakeholder responded to every question. The questions were open-ended and thus issues raised by stakeholders had to be "volunteered." We did not ask each stakeholder to agree or disagree with particular issues.

[39] OPP officials noted that they share these concerns, but they stated that the results of OPP's internal review of conditional registrations convinced them that these concerns have no basis.

[40] EPA officials noted that there is no legal requirement for EPA to acknowledge receipt of required information or notify registrants whether the data submitted satisfy the condition imposed.

[41] EPA officials noted that the public interest finding only applies to FIFRA section 3(c)(7)(C) related to conditional registration of pesticide products with new active ingredients, and not to sections 3(c)(7)(A) or (B) related to conditional registration of products that are identical or substantially similar to products already registered or new uses of products already registered, respectively.

[42] Specialty crops include fruits and vegetables, tree nuts, dried fruits, and nursery crops (e.g., flowers).

[43] EPA officials stated that EPA must have all of the data pertaining to a new use of a pesticide before approving the use. Thus, according to these officials, EPA fully evaluates the human health and environmental impacts of adding the new use.

[44] As of January 2013, EPA officials noted that the agency is considering a request for suspension of this product. EPA denied a previous request to suspend this product. According to these officials, the current suspension request is based on a dispute related to the science underlying the registration, not the fact that the registration is conditional per se. EPA

officials also noted that because EPA is involved in active litigation on Clothianidin, they are unable to respond to these stakeholder comments.

[45] EPA officials noted that the additional data it required for the conditional registration of Imprelis were related to human health, specifically dietary/food exposure. According to these officials, EPA received required studies from the company related to nontarget plants. EPA reviewed these studies, found no concerns, and the product was conditionally approved. However, after the product was marketed, it became apparent that it was killing trees. According to EPA officials, based on the studies the company submitted, there was no reason to suspect that the product would impact trees. However, a study specifically for effects on trees was neither required by EPA nor submitted by the company. EPA officials stated that the tree deaths were very unfortunate and made the product unacceptable and that if EPA had known about the impacts on trees at the time the product was being considered for registration, the agency might not have registered the product at all. EPA officials stated that this was not a case where the agency failed to anticipate risks, rather, it was a risk that the standard data requirements did not capture and a risk that only came to light after the product was registered and marketed.

[46] GAO, *Nanotechnology: Nanomaterials Are Widely Used in Commerce, but EPA Faces Challenges in Regulating Risk,* GAO-10-549 (Washington, D.C.: May 25, 2010).

[47] EPA has previously addressed toxicity concerns regarding nanosilver. In 2010, EPA issued its *State of the Science Literature Review: Everything Nanosilver and More* in which EPA reported that there is clear evidence that nanosilver is toxic to aquatic and terrestrial organisms and may be detrimental to human health. In this report, EPA noted that several studies have shown that nanosilver can be released into the wastewater stream during washing, such as from socks containing nanosilver. The nanosilver released may disrupt the helpful bacteria used in wastewater treatment processes or be released into the environment. 48EPA officials said that the basis of the Natural Resources Defense Council lawsuit is not a challenge of EPA's use of conditional registrations under FIFRA section 3(c)(7) or whether EPA met the standards for granting a conditional registration under this statute. Rather, according to these officials, the lawsuit questions an aspect of the risk assessment methodology that EPA used in this case to make the determination of no unreasonable adverse effects.

[49] EPA Office of the Inspector General, EPA Needs to Manage Nanomaterial Risks More Effectively, Report no. 12-P-0162, Dec. 29, 2011. OPP officials noted that the Office of the Inspector General advised EPA to develop a better internal process for sharing data across program offices. According to these officials, EPA has implemented this recommendation, and an Inspector General official concurred.

[50] National Research Council of the National Academies, A Research Strategy for Environmental, Health, and Safety Aspects of Engineered Nanomaterials, the National Academies Press (Washington, D.C.: 2012).

[51] These statutes and regulations include Federal Insecticide, Fungicide, and Rodenticide Act, Act of June 25, 1947, ch. 125, 61 Stat. 163 (codified as amended at 7 U.S.C. §§ 136-136y); Federal Food, Drug, and Cosmetic Act, 21 U.S.C. §§ 301-399f; Food Quality Protection Act of 1996, Pub. L. No. 104-170 110 Stat. 1489; Pesticide Registration Improvement Act of 2003, Pub. L. No. 108-199 Div. G. Tit. V. § 501. 118 Stat. 419 (2004); and relevant parts from title 20 of the Code of Federal Regulations.

[52] These program and guidance documents included EPA, Conditional Registrations – Summary and Analysis, Office of Pesticide Program Internal Review, Spring 2011; Pesticide Programs: Conditional Registration of New Pesticides, F.R. Vol. 51, No. 43, p.p. 7628-7634; and EPA Biological and Economic Analysis Division (BEAD), Guidance for BEAD Review: Public Interest Findings for Conditional Registration.

[53] GAO, Internal Control Management and Evaluation Tool, GAO-01-1008G (Washington, D.C.: August 2001); GAO/AIMD-00-21.3.1; Office of Management and Budget, Management's Responsibility for Internal Control, OMB Circular No. A-123, Dec. 21, 2004; Office of

Management and Budget, Managing Government Records Directive, M-12-18, Aug. 24, 2012.

[54] Previous GAO reports reviewed are listed in "Related GAO Products" at the end of this report. Previous EPA Inspector General reports include: EPA Office of Inspector General, EPA Needs to Manage Nanomaterial Risks More Effectively, Report no. 12-P-0162, Dec. 29, 2011; EPA Office of Inspector General, EPA Needs to Comply with the Federal Insecticide, Fungicide, and Rodenticide Act and Improve Its Oversight of Exported Never-Registered Pesticides, Report no. 10-P-0026, Nov. 10, 2009; EPA Office of Inspector General, Office of Inspector General Semiannual Report to the Congress, Report no. EPA-350-K-00-002, November 2000; EPA Office of Inspector General, EPA's Pesticide Program, Report no. E1EPE2-05-0015-4100205, Mar. 11, 1994.

[55] These key changes followed implementation of the Food Quality Protection Act of 1996.

[56] Because this was a nonprobability sample, the information collected in response to the questionnaire about EPA's use of conditional registrations cannot be generalized to all consumer, environmental, industry, legal, producer, science, and state government stakeholders but provides illustrative information on the views of such stakeholders by category.

[57] Seven stakeholders did not respond to our request to complete the questionnaire. Although three stakeholders provided complete responses, we did not consider them valid because they responded as individuals instead of as representatives of their respective organizations. Another stakeholder declined to fill out the questionnaire.

INDEX

H

I

J

K

L

M